D0022717

Introduction to Statistics

Introduction

Second Edition

HOUGHTON MIFFLIN COMPANY **Boston**

Atlanta

Dallas

Geneva, Illinois

Hopewell, New Jersey

Palo Alto

London

to Statistics

A NONPARAMETRIC APPROACH

Gottfried E. Noether
University of Connecticut

Printed in the U.S.A.
Library of Congress Catalog Card Number: 75-19532
ISBN: 0-395-18578-5

To L.

CONTENTS

EDITOR'S INTRODUCTION TO FIRST EDITION

The introductory course in statistics has always represented a serious pedagogic problem. While it will be terminal for many students, for others it will provide a basis for studying specialized methods within their major fields. The main function of such a course is to introduce them to variability, uncertainty, and some common statistical methods of drawing inferences from observed data. Such a course should be intellectually stimulating; the ability to reason statistically should be more crucial to the success of a student than his mathematical ability.

Traditional textbooks do not generally serve the needs adequately. In most cases the basic statistical ideas do not appear until very late. Valuable time is first devoted to the intellectually dull task of organizing and graphing data, and to interminable computations of means, modes, medians, and variances. Sometimes effort is still devoted to rather pointless discussions of skewness, kurtosis, and geometric means. Students with little interest in mathematics find a week or more of combinatorics frustrating if not mystifying. Finally one-quarter or one-semester courses which attempt to cover most of the important methods in statistics run out of time or become dull compendia of formulae. It is preferable to concentrate on basic ideas and a few methods and to let the applied departments introduce specialized techniques when needed at a small cost of time.

What are called for are innovative approaches which avoid these pitfalls and are efficient in exhibiting many basic ideas of inference in contexts which are intellectually stimulating. Professor Noether has presented us with such a book. The main technique is the exploitation of nonparametric methods. This book is not a study of nonparametric methods in statistics. Rather, it is the effective use of these methods to illustrate statistical ideas.

What are the advantages of this approach? The basic ideas of inference appear at the beginning. The methods developed are easy to apply and require a minimal amount of computation. The methods are simple in principle; the common sense logic behind them is easy to perceive and to explain. The probability background used in most of the illustrations is minimal, requiring often no more than easy applications of the binomial distribution and its tables. Although no attempt is made at a complete coverage, this vehicle is remarkably effective in covering many basic ideas with a few methods and illustrations. Finally, there are the advantages of the robustness and of the wide applicability of the nonparametric methods discussed.

This is a well-conceived, carefully thought out, and well-written book. I fully anticipate that Professor Noether's success with a preliminary version will be shared by many teachers who use this book.

Herman Chernoff

PREFACE TO SECOND EDITION

A first course in statistics for nonmajors poses diverse problems. Students are easily bored by endless and to them often meaningless computations. At the other extreme, they are frequently overwhelmed by too much probability theory. In this book topics on probability theory and on descriptive aspects of statistics are held to bare essentials. Instead, the book concentrates on basic ideas without getting overly involved in computational or technical detail.

Estimation and hypothesis testing are discussed in terms of the binomial model which is much simpler than the customary normal model. One- , two- , and k-sample problems are solved nonparametrically before the student is introduced to t-tests and the analysis of variance. Testing for randomness against the alternative of a monotone trend and measuring strength of relationship through rank correlation precedes least-squares regression and product-moment correlation. This arrangement is based on my conviction and experience that nonparametric methods are not only safer but are also conceptually and technically simpler than normal theory methods. At the elementary level, nonparametric methods require much less preparation in probability theory than normal theory methods. All that is needed is an understanding of equally likely cases. If the instructor so desires, sampling distributions of relevant test statistics can be obtained by simple enumeration. There is no need to spend valuable time on discussion of means and variances of sums of random variables.

The overall organization of the Second Edition is unchanged (except that the chapter on random variables is now placed in an appendix). Many chapters have been extensively rewritten. In the chapters on hypothesis testing, one-sided tests are now taken up before two-sided tests; type 2 errors are discussed simultaneously with type 1 errors; and descriptive levels are stressed more strongly than before. In the chapter on contingency table analysis, tests of homogeneity including the important 2×2 case are taken up before tests of independence. The Kendall rank correlation coefficient now emerges naturally as an extension of the test of randomness against the alternative of a monotone trend.

The chapters dealing with normal theory methods have been greatly simplified and tied in more closely with the corresponding nonparametric discussion. Instructors who feel that a one-semester course is incomplete if it does not cover the one-sample t-test should find it quite feasible to go from the sign test in Chapter 12 directly to the t-test in Chapter 17.

I have taken the advice of the Panel on Statistics of the Committee on the Undergraduate Program in Mathematics and deemphasized topics such as the use of the continuity correction and tests involving variances. Except for the table of normal probabilities, tables have not been changed. In response to repeated requests, a table of square roots has been added, and the number of problems has been substantially increased. The author gratefully acknowledges the four volumes *Statistics by Example* prepared by the Joint Committee on the Curriculum in Statistics and Probability of the American Statistical Association and the National Council of Teachers of Mathematics as the inspiration for several of the new problems.

The book is intended for a one-semester or two-quarter course. I usually cover most of the material in Chapters 1 to 14 and Chapter 16 in a one-semester course meeting three hours a week. Other possibilities are indicated in the schematic outline on page xv. Optional sections are indicated by an asterisk. While these sections broaden the scope of the discussion and sometimes fill in mathematical details, they can be omitted from classroom discussion without loss of continuity. A reasonably good high school background in mathematics is sufficient for the book.

I express my deep appreciation to Herman Chernoff, Ralph D'Agostino, and John Pratt for many helpful suggestions in connection with the First Edition and to Thomas Hettmansperger and Ralph Mansfield in connection with the Second Edition. I am thankful to Robert Stephenson for checking computational details and preparing answers for all the problems. Sally Postemsky did a marvelous job typing the manuscript for the Second Edition.

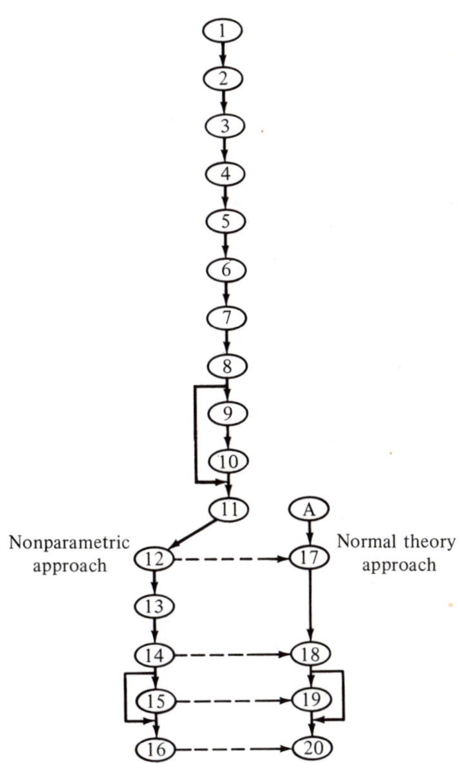
1
2
3
4
5
6
7
8
9
10
11
A
Nonparametric approach
12
17
Normal theory approach
13
14
18
15
19
16
20

1

WHAT IS STATISTICS?

If experimentation is the Queen of the Sciences, surely statistical methods must be regarded as the Guardian of the Royal Virtue.

(From a letter to *Science* by Myron Tribus)

1 To most people the word "statistics" brings visions of long columns of numbers, mystifying graphs, and frightening charts that show how the government is spending our tax money. At one time the word referred exclusively to numerical information required by governments to conduct state business. Statisticians were people who collected masses of numerical information. Some statisticians still function in this way, but many do not. They help in the conduct and interpretation of scientific experiments and professional investigations. Along with changes in the work of statisticians has gone a change in the meaning of the word "statistics" itself. Statistics may refer to numerical information, like football statistics or financial statistics. However, the word can also refer to statistics as a discrete discipline like mathematics or economics. When we use the word "statistics" in our discussions, we will usually have this second meaning in mind. The few times the word refers to numerical information as such, the fact will be clear from the context.

How, then, can we describe the field of statistics? A United States Civil Service Commission document says that "statistics is the science of the collection, classification, and measured evaluation of facts as a basis for inference. It is a body of techniques for acquiring accurate knowledge from incomplete information; a scientific system for the collection, organization, analysis, interpretation and presentation of information which can be stated in numerical form." I just hope that this definition does not keep anyone from wanting to study statistics!

The following is a much simpler statement: Statistics deals with ideas and methods to improve the drawing of conclusions from numerical information in the face of uncertainty. While it may not reflect so precise and all-inclusive a view of statistics as the earlier statement, it emphasizes the aspect of statistics that is of principal interest to us in this book, namely, how to use incomplete information to draw valid conclusions and make useful decisions.

The kind of decisions that concern a statistician are decisions based on numerical information. All of us are accustomed to making decisions of this type. For example, when we have to drive to the airport and wonder how much time to allow for the trip, we try to generalize from past observations.

Often past experience coupled with common sense produces satisfactory results. On the other hand, there are many problems of a statistical nature for which common sense alone fails to provide adequate answers. What we need are more formal methods—methods of statistical inference as they are usually called—developed and analyzed by mathematical statisticians using the calculus of probability. Starting with Chapter 6, we will study some of the simpler inference methods and the rationale behind them. In Chapters 2–5 we will learn some basic probability. In the remainder of this chapter we will illustrate certain ideas that are basic for statistical inference. Of necessity, this preliminary discussion will rely heavily on intuition.

The Taxi Problem

2 Suppose that you are waiting at a street corner for a taxi. Several drive by, but they are all occupied. You begin to wonder how many taxis there are in this city. Clearly there are not enough to go around. But how many are there? You start watching the numbers on the shields of the taxis going by:

405, 280, 73, 440, 179.

The next taxi stops for you. You could simply ask the taxi driver how many taxis are registered in the city. But before you do so, you wonder whether you can make an educated guess on the basis of your observations. Indeed you can. As a starter, let us represent our information graphically by marking the five observed numbers on a line as follows:

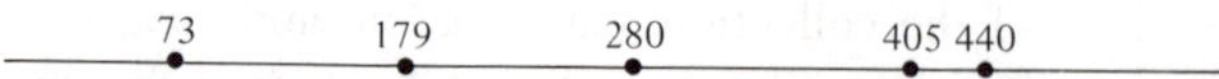

The endpoint on the left corresponds to taxi #1. We should like to know where to put the endpoint on the right. How can we make a reasonable guess?

We might argue as follows. We do not know how many taxis have numbers that are greater than the largest observed taxi number. What we do know is that there are 72 taxis with numbers that are smaller than the smallest observed taxi number. The following modified graph illustrates the situation:

Suppose that we replace the question mark on the right by the corresponding number on the left, namely 72. Our guess then becomes $440 + 72 = 512$. In statistical terminology we refer to 512 as an estimate of the total number of cabs. Our particular estimate is obtained by adding one less than the smallest observed taxi number to the largest observed taxi number.

Of course, we are curious to know how close our estimate comes to the true value. The taxi driver informs us that there are 550 taxis in the city. Thus our estimate is in error by 38.

The type of investigation we have just described is known as *serial number analysis*. One interesting application of serial number analysis occurred during World War II, when the Allies wanted to know how many German tanks there were. When the war was over, it was discovered that estimates of German tank production based on serial number analysis had been much more accurate than estimates derived from more orthodox intelligence sources. The method used to estimate German tank production was somewhat more sophisticated than our approach in the taxi case, but it was based on similar reasoning.

3 We stated earlier that we wanted to illustrate certain basic ideas of statistical inference. Statistical inference is concerned with drawing useful conclusions from available data, usually called the *sample*, about a larger aggregate, called the *population*. In our taxi example the sample consists of the taxis numbered 73, 179, 280, 405, and 440 that we happened to observe. The population consists of all taxis from 1 to 550. In this example the quantity that has the greatest interest is the number 550. Before the taxi driver told us this number, it was unknown to us. But by looking at the five numbers in our sample, we were able to make a rather good guess.

A different kind of problem arises if we have some preconceived idea of what the total number of taxis might be. Somebody might have assured us that there are at least 1000 taxis in the city. After waiting a long time for a taxi to pick us up and noting that all passing taxis have numbers smaller than 500, we may develop some doubts about the correctness of the claim. Formally, we set up a statistical hypothesis and test it on the basis of experimental evidence. The hypothesis states that the number of taxis in the city is at least 1000. The testing process consists of making an evaluation of how well our experimental evidence can be reconciled with the stated hypothesis. In Section 22 in Chapter 3 we set up a mathematical model that allows us to compute certain relevant probabilities. As we will see, these probabilities suggest that the claim that there are at least 1000 taxis around is hardly tenable.

We can indicate the nature of the argument without getting too deeply involved in probability computations by comparing the experiment of observing five taxi numbers with an experiment involving five tosses of a coin. In the taxi experiment none of the taxi numbers exceeded 500 (indeed, none exceeded 440). We might compare this result with the event of observing tails five times in a row when tossing a coin. A captain of a football team would very likely feel aggrieved if he lost the toss five times in a row. After all, there is only 1 chance in 32 of such an event happening by chance. (There is 1 chance in 4 of losing the toss twice in a row; there is 1 chance in 8 of losing the toss three times in a row; etc.) Similarly, if a city has 1000 taxis numbered from 1 to 1000, there is only 1 chance in 32 that, due strictly to chance, none of five observed taxis have numbers exceeding 500. (The probability that none exceeds 440 is considerably smaller, namely .016, as we will see in Section 22.) However, while our football captain can only blame Lady Luck for his misfortune (unless he is prepared to accuse his opponents of cheating, a possibility we will consider in Chapter 7), explanations based on other than pure chance considerations are possible in the taxi example. Our probability computation is based on the assumption that 1000 taxis are around. If, for example, the city has only 600 taxis, there is nothing unusual about each of five observed taxis having a number of 500 or lower.

We may summarize the situation as follows. The original claim of 1000 taxis is not necessarily false, but if it is correct, then we have witnessed an event that has an exceedingly small probability. Events with small probabilities do occur. There are football captains who lose the toss five or even more times in a row and simply blame Lady Luck. But rather than ascribe matters to chance in statistical investigations, statisticians prefer to look for alternative explanations. In the taxi example a more reasonable reaction is to reject the claim that there are at least 1000 taxis around and settle for a more modest number. The estimate of Section 2 suggests that there are not much more than 500 taxis around.

4 Estimation and hypothesis testing are the two main topics we will discuss in this course. As we will see, there are many problems associated with estimation and testing. Another look at taxi estimation calls attention to one such problem. The method of estimating the total number of taxis proposed in Section 2 is not the only possible method. If we do come up with two (or more) estimates of the same quantity using different methods, which estimate should we use? Let us be more precise and actually propose a second estimate for the total number of taxis.

Rather than concentrate on the largest and smallest taxi numbers in the sample as we did with our first estimate, let us concentrate on the number "in the middle" of the sample, the so-called median. This time we assume that the number of taxis having numbers greater than the median equals the number of taxis having numbers smaller than the median. Since the median of our five taxi numbers equals 280, we have the following picture:

New estimate
280
279
279

The new estimate is $280 + 279 = 559$.

In our example the second estimate is clearly better than the first estimate, since 559 is closer to the true value, 550, than is 512. In general, however, there is no taxi driver around to tell us the true value after we have computed our different estimates. The Germans certainly did not tell the Allies how many tanks they had built during a given period of time. The Allies found out only at the end of the war, when the information was purely academic. In most statistical problems we never learn the true value.

How can we decide which estimate to use, the first or the second? Before we give an answer to this question, let us look at some additional data, representing three more sets of five observations of taxi numbers. These numbers were taken from tables of random numbers rather than from actual taxicab observations. We will hear a great deal more about random numbers later on. At this moment all we have to know is that the results are comparable to an experiment in which three statisticians stand at three busy street corners in a city with 550 taxis numbered from 1 to 550, each writing down the numbers of the first five taxis that happen to pass by. The numbers in our original sample as well as those in the additional three samples are listed in Table 1.1.

TABLE 1.1

Sample 1	405	280	73	440	179
Sample 2	72	132	189	314	290
Sample 3	191	124	460	256	401
Sample 4	450	485	56	383	399

Our earlier discussion shows that the first estimate equals $L + S - 1$, where L denotes the largest observation in the sample, and S, the smallest. If M stands for the sample median—that is, the number in the middle after observations have been arranged according to size—the second estimate is given by $M + (M - 1)$ or more simply, $2M - 1$. We then find the estimates in Table 1.2, where we use the names extreme estimate and median estimate to refer to the first and second estimates, respectively. The numbers in parentheses indicate by how much each estimate is in error, that is, by how much it differs from 550.

TABLE 1.2

	Extreme Estimate	*Median Estimate*
Sample 1	512 (38)	559 (9)
Sample 2	385 (165)	377 (173)
Sample 3	583 (33)	511 (39)
Sample 4	540 (10)	797 (247)

One thing is clear from our computations: Neither estimate is consistently better than the other. For the first sample, the median estimate is better than the extreme estimate, as we have already noted. For the second and third samples, the extreme estimate is slightly better than the median estimate, and for the fourth sample, the extreme estimate is overwhelmingly better. On the basis of the available information, a clear-cut decision as to which estimate is better is not possible. However, by continuing the experiment, we would eventually find that, on the average, the extreme estimate is closer to the true value than the median estimate. Some indication of this is contained in Table 1.2. The average error for the four median estimates is $(9 + 173 + 39 + 247)/4 = 117$; the average error for the corresponding four extreme estimates is considerably smaller, namely, $(38 + 165 + 33 + 10)/4 = 61.5$. This is where the mathematical statistician comes in. With the help of the theory of probability, a mathematical statistician can not only determine which of two estimates is better, but can also determine how much better one estimate is than the other. However, such investigations are beyond the scope of our course. In general, we have to take the word of a mathematical statistician that a recommended procedure has desirable properties.

Assumptions

5 The previous discussion has been rather intuitive. However, it is important to realize that underlying our intuitive approach, there are various assumptions without which the suggested solutions would have very doubtful value. For example, we assumed that taxis in the given city were numbered consecutively starting with 1. This seems like a reasonable assumption but need not necessarily be true. For instance, in the corresponding problem of estimating German tank production, it soon became evident that German tanks were not numbered in any one consecutive sequence, complicating the estimation problem considerably.

Nearer to home the author conducted the following experiment. In one of the Boston suburbs, he noted the numbers of the first five taxis that crossed a certain main intersection and showed evidence of being registered locally with the following results:

$$35, 18, 38, 43, 23.$$

The extreme estimate becomes $43 + 18 - 1 = 60$. Inquiry revealed that there were actually only 40 taxis registered in town. The estimate was in error by 50 percent. The reasons is easy to see. Since the largest observed number was actually 43, not all numbers could have been in use. Indeed, even a cursory inspection of the numbers actually observed reveals an absence of low numbers. Apparently, all or most taxis with lower numbers had been taken out of use.

A second assumption that is more difficult to check is the equal likelihood of each taxi passing the corner where we are waiting to be picked up. This assumption would not be appropriate if, for example, individual taxis are assigned to different sections of the city.

It is important to realize that statistical conclusions are not absolute. They are valid only within a given framework. The more realistic a framework we choose, the more reliable our conclusions will be. A given statistical method may be appropriate within one framework but inadequate within another. An important part of a statistician's work is to select a reasonable framework—or probability model, as we will usually say—and then use methods of statistical analysis that are appropriate for this model. A statistical technique that is based on assumptions of a relatively general nature is often preferable to a more elaborate technique that assumes a more restrictive framework. As we will see later on, the nonparametric methods discussed in Chapters 12–16 have very general validity, while the corresponding methods in Chapters 17–20 assume a more restrictive framework.

Medians

6 One of the previously discussed estimates requires the determination of the median M of a set of numbers. Since the median will be used repeatedly in other statistical problems presented in subsequent chapters, we close this chapter with a discussion of its computation.

In our examples the computation of the median was easy. Among five observations, all different, the median is simply the observation "in the middle," that is, the third largest (or smallest) among all five observations. However, this description is not sufficiently precise to determine the median of an arbitrary set of numbers. We need a general definition of the median and a simple procedure for determining it in practice. The median M of a set of numbers has the following property: There are (as nearly as possible) as many numbers that are smaller than M as there are numbers that are greater than M. More specifically, the median of an odd number of observations is exactly the middle item when numbers are arranged from the smallest to the largest. When the number of observations is even, in general, no single middle item exists. So we take the average of the middle two items as the median.

EXAMPLE 1.3 The median of 405, 280, 73, 440, and 179 is 280.
The median of 405, 280, 73, and 440 is $(280 + 405)/2 = 342.5$.
The median of 405, 280, 440, 179, and 405 is 405. (In this example the third and fourth largest numbers are both equal to 405. Since by definition the median M equals the third largest, $M = 405$. In this example two numbers are smaller than M; only one number is greater than M.)
The median of 405, 280, 405, and 405 is 405. (Both the second and third smallest numbers equal 405.)

As long as we do not have too many numbers, a convenient method for determining the median M is as follows. We cross out the largest and the smallest numbers and continue crossing out numbers in pairs until either one or two numbers are left. At each step the largest and smallest numbers that have not been previously eliminated are crossed out. If a single number is left at the last step, this number is M. If two numbers are left, M is chosen as the number halfway between the two numbers. Any number that occurs more than once is used in this crossing-out process as often as it occurs.

EXAMPLE 1.4 For the set 405, 280, 73, 440, 179, and 405, we start by crossing out 440 and 73. This leaves 405, 280, 179, and 405. Now we cross out 405 and 179, leaving 280 and 405. Thus $M = (280 + 405)/2 = 342.5$.

PROBLEMS

1 Find the median for each of the following sets of numbers:

a. 27, 13, 14;
b. 35, 49, 68, 50;
c. 25, 25, 25;
d. 46, 26, 39, 45, 26, 11, 51;
e. the 20 taxi numbers in Table 1.1.

2 Find the median for the data in Problem 5 of Chapter 12.

3 In Table 14.1 find the median of the

a. x-scores;
b. y-scores; and
c. x- and y-scores combined.

4 Using the data in Problem 15 of Chapter 14, find the median income of families in

a. community A;
b. community B;
c. communities A and B combined.
d. Using your intuition, how would you answer part a of Problem 15, Chapter 14?

5 In a certain community, in a survey of 100 families the following number of children were reported:

Number of children:	0	1	2	3	4	5	*more than* 5
Number of families:	17	20	28	19	7	4	5

Find the median number of children for these families.

6 Check the extreme and median estimates for Samples 2 through 4 in Table 1.2.

7 Find the extreme and median estimates for the following two sets of taxi numbers:

a. 309, 769, 78, 61, 277, 188;
b. 209, 181, 595, 799, 694, 334, 595.

8 Consider the four samples of size 5 in Table 1.1 as one sample of size 20. Find the extreme and median estimates.

9 A glass bowl contains tags numbered from 1 to t. Three tags are selected in a random fashion. Their numbers are 35, 63, and 26. Find the extreme and median estimates of t.

10 At a supermarket the shopping carts bear number tags. You are standing in line at the checkout counter and note the following four number tags: 65, 88, 68, and 16. What is your estimate of the total number of shopping carts in the store?

11 On the Mexican island of Cozumel, telephones have consecutive numbers starting with 1. On arrival a visitor calls three hotels having numbers 117, 72, and 137. What is your estimate of the total number of phones on Cozumel? Do you think that this is a very reliable estimate? Why?

12 While standing in line at a chairlift, a skier notices that the chairs have number tags. The six skiers in front of him have chairs with number tags 5, 36, 23, 29, 27, and 41.

a. What is your estimate of the total number of chairs?
b. If the skier's own chair has tag number 14, how does this affect the earlier estimate? (Answer for both the extreme and median estimates.)

13 Compare the *median* error of the four extreme estimates in Table 1.2 with the *median* error of the four median estimates. What does this comparison suggest?

14 Given a set of numbers, the difference between the largest and smallest numbers in the set is called its *range.*

a. Find the range of the four extreme estimates in Table 1.2.
b. Find the range of the four median estimates in Table 1.2.
c. On the basis of the ranges, which estimate seems preferable, the extreme estimate or the median estimate? Why?

In Problems 15 and 16 let the symbol t stand for the total number of taxis.

15 Get a sample of ten taxi numbers for your city. Estimate t using the extreme estimate. Try to find out the true value of t for comparison. If your estimate is in error by, say, more than 20 percent, try to find reasons.

16 Suggest one or two other methods for estimating the number t. Compute the corresponding estimates for the four samples in Table 1.1. Do they seem to be better or worse than the extreme estimate? Than the median estimate?

17 Assume that the fifth taxi number in Sample 2 of Table 1.1 is 390 (rather than 290). Recompute the extreme and median estimates. Why does this example show that the median estimate may produce rather undesirable estimates?

18 A slightly more complex (but also better) estimate than the extreme estimate is the following *gap* estimate. Using again the example of Section 2, we compute all possible gaps (that is, the number of unobserved taxis).

73 179 280 405 440

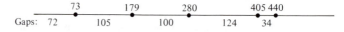

Gaps: 72 105 100 124 34

The extreme estimate is found by adding the gap farthest to the left to the largest observation in the sample. The gap estimate is found by adding the average of all gaps to the largest observation. Thus

$$\text{Gap estimate} = 440 + \frac{72 + 105 + 100 + 124 + 34}{5} = 440 + \frac{435}{5}$$
$$= 440 + 87 = 527.$$

a. Find the gap estimates for the samples in Table 1.1.
b. Find the average error for the four estimates.

19* Show that the gap estimate of Problem 18 can be computed by means of the formula

*An asterisk beside certain problem numbers throughout the book indicates that the problem has primarily theoretical interest and/or refers to a section that has an asterisk.

$$\text{Gap estimate} = \frac{n+1}{n}L - 1,$$

where n is the number of observations in the sample and L is the largest observation. (*Hint:* The sum of all n gaps equals $L - n$.)

20 How would you estimate the number of fish in a lake? Fish do not carry serial numbers, but statisticians have devised the so-called *capture–recapture* method to solve this and similar problems. A first group of fish is caught, tagged, and released. A few days later a second group of fish is caught, and the statistician observes how many of the fish in the second group have tags. As an example, assume that the first group contains 50, and the second group contains 100 fish, and that 5 fish in the second group have tags. Can you make a reasonable estimate of the total number of fish in the lake? (*Hint:* 5 percent of the fish in the second catch have tags. Assume that the same is true of all fish in the lake.)

THE MEANING OF PROBABILITY

The Frequency Interpretation

7 In Chapter 1 we observed that probability considerations form the basis of statistical inference. In this and the next chapter we will discuss some simple aspects of probability.

We begin by introducing some notation. Probabilities are associated with *events,* like the event that it rains, the event of twin births, the event that on the next airplane trip our luggage will go astray. We will use letters like A, B, and so on, to denote events. For the probability of the event A, we write $P(A)$ and read "the probability of A" or more briefly, "P of A." As an example, in books on card games we read that when we deal five cards from a well-shuffled deck of playing cards, the probability is $^{198}/_{4165}$ that the five cards contain "two pairs" (for instance, 2 kings, 2 tens, and an ace). If we denote the event "two pairs" by the letter T, we would write $P(T) = {}^{198}/_{4165}$. Probabilities like this one are not difficult to compute, provided that we know some basic principles of what the mathematician calls *combinatorial analysis,* but right now we are more interested in knowing what such a probability means. We note that $^{198}/_{4165}$ does not differ very much from $^{200}/_{4000} = {}^{1}/_{20}$; so, for simplicity, let us discuss what is meant by the statement that a certain event has probability $^{1}/_{20}$. For many practicing statisticians, events having probability $^{1}/_{20}$ constitute a sort of bench mark in hypothesis testing.

When we deal five cards from a deck of cards, they either contain two pairs or they do not. There are no other possibilities. Where does the $^{1}/_{20}$ come in? The answer to this question can be found in certain empirical facts that were noted by gamblers hundreds of years ago. Consider any game of chance. If the game is played honestly, we cannot predict the outcome of a single game, or *trial*, as we will often say. But something rather remarkable happens when we play the same game over and over again. If we look only at the outcomes of successive trials, no pattern emerges. However, the relative frequency with which any given outcome occurs shows greater and greater regularity and eventually becomes nearly constant. When we talk of the probability of an event, we have this limiting value of the relative frequency in mind.

Table 2.1 illustrates this phenomenon. The data represent a partial record of what happened when a "coin" was tossed 5000 times. We have recorded how often heads occurred during the first 10 tosses, the first 20 tosses, and so on. By dividing the number of trials, we obtain the relative frequency of heads.

TABLE 2.1 **Results of 5000 Coin Tosses**

Number of Trials	*Number of Heads*	*Relative Frequency*
10	7	.700
20	11	.550
40	17	.425
60	24	.400
80	34	.425
100	47	.470
200	92	.460
400	204	.510
600	305	.508
800	404	.505
1000	492	.492
2000	1010	.505
3000	1530	.510
4000	2030	.508
5000	2517	.503

The sequence of relative frequencies holds the greatest interest for us. During the early stages of the experiment, the relative frequencies show considerable variability, but after hundreds and thousands of trials they become more nearly constant. Presumably, if the experiment were continued, there would be less and less fluctuation in the relative frequency, suggesting eventual convergence to a constant, which would then be called the probability of obtaining heads in a single toss of a coin, or more briefly, the probability of heads.

8 What we have just discussed is, for obvious reasons, called the *frequency interpretation of probability*. For our purposes, the frequency interpretation is extremely useful. However, other interpretations of probability are also possible. There are situations where the notion of unlimited repetition of an experiment as required by the frequency

interpretation is, to say the least, very artificial. For example, people talk of the probability that the Boston Red Sox will win the pennant this coming year. Here we are interested in one particular pennant race. It is possible to argue that the same is true of the toss of a coin. However, it is easy for us to think of the toss of a coin as one of a long sequence of similar tosses. The same cannot be said of the American League pennant race with its ever-changing conditions. With an event like a pennant race, we then often speak of *personal* or *subjective* probabilities. Such probabilities give numerical expression to a person's *intensity of belief*, that is, his conceivable willingness to bet a certain amount of money on the occurrence or nonoccurrence of a given event.

From a purely mathematical point of view, personal probabilities can be dealt with in the same way as probabilities that have a frequency interpretation. They can be, and often are, the basis of decisions in the face of uncertainty. However, when we talk of probabilities, we will usually have a frequency interpretation in mind.

The eventual constancy of the relative frequency of events associated with repetitive experiments is not a property of games of chance alone. While the theory of probability obtained its start from problems raised by gamblers in the seventeenth century, its spectacular growth, particularly during the last thirty or forty years, is due to the realization that the methods that give answers to questions posed by gamblers also answer questions of scientists, engineers, and business people. Indeed, it is difficult to imagine a field of human endeavor in which probability is not used in one way or another. One of its most important applications is in the field of statistics.

To repeat what we have learned about the frequency interpretation of a given probability: When we say that the probability of a given event is, for example, $1/20$, we imply that the relative frequency with which the event would occur in a long sequence of trials eventually stabilizes at the value $1/20$. It is customary to express this result by saying that we expect the event to happen about once in every twenty trials, or about five times in one hundred trials. There is no harm in such a statement if we remember its correct meaning in terms of what happens in the long run and do not interpret it literally. The statement does not mean, as many people seem to think, that if an event with probability $1/20$ has not occurred in nineteen successive trials, it is bound to occur on the twentieth trial. On the twentieth trial, the event still has only one chance in twenty of occurring.

Random Numbers

9 We conclude this chapter with a brief discussion of *random digits*, or *random numbers*, as they are often called. Some students may have wondered about the use of quotation marks in the description of the

coin-tossing experiment. The explanation is that no coin was used in collecting the data. Real coin tosses take a great deal of time and require careful attention to keeping conditions unchanged from one toss to another. Instead, the basic information came from a *table of random digits.* We have mentioned a table of random digits once before in our discussion of the taxicab problem. What then is a table of random digits?

Let us consider the following experiment. We have 10 identical pingpong balls on which we have written the digits 0, 1, . . . , 9. After mixing the balls very thoroughly in a box, we select one of the balls without looking, note down the digit written on the ball, and then put it back in the box. We repeat the whole process of mixing, selecting, noting down a digit, and returning the ball to the box again and again. The result will look something like this:

40582 00725 69011 25976 .....,

where the digits have been written in groups of five for easier reading. A table of random digits is simply a more extensive listing, which is usually produced by an electronic computer instead of by an actual experiment like the one we just described.

In Chapter 3 we will discuss random digits in greater detail. Now we are content to observe that random digits have the property that each of the 10 digits 0 through 9 has probability $\frac{1}{10}$ of occurring in any position in a random-number table.

Random digits can be used to simulate a coin-tossing experiment like the one described earlier. Each digit in a sequence of random digits may be assumed to furnish the result of a coin toss by, for example, letting even digits correspond to heads and odd digits, to tails. In this way the first 10 digits in our listing produce 7 heads and 3 tails; all 20 digits produce 11 heads and 9 tails. (These are the first 20 "tosses" in Table 2.1.)

Random digits play an important role in many statistical investigations. Many high-speed computers have built-in provisions for producing random digits as needed. Printed tables are also available the largest of which contains 1000000 digits. The 5000 digits that formed the basis of our coin-tossing experiment were taken from this table. Table J at the end of the book is a short table of random numbers.

PROBLEMS

For Problems 1, 3, and 5, students may want to work together in small groups.

1 Roll a die 120 times and observe how often it falls 1, 2, . . . , 6. Are the results in agreement with what you would have expected in such an experiment? (Keep your data for later use.)

2 Imagine an experiment in which you roll two dice 180 times and observe how often the two dice add up to 2, 3, . . . , 12. Invent appropriate data for such an experiment. (Be sure to do this problem before doing Problem 3. Keep the results.)

3 Actually perform the experiment described in the preceding problem. Are the results similar to those for Problem 2? (Keep the data.)

4 Imagine the same kind of experiment as in Problem 1, but involving 6000 rolls. Invent appropriate data and keep them for later use.

5 Select 250 telephone numbers from a telephone book. Using only the last four digits in each number—that is, ignoring prefixes—tabulate the frequency of 0s, 1s, . . . , 9s. Do all ten digits seem to appear with equal frequency? (Keep the data.)

6 Group Project (for groups of 50 or more students): Each student in a group writes down a digit between 0 and 9 on a slip of paper. Collect all slips and determine the proportions of 0s, 1s, ... (Keep the data.)

7 Students should collect and keep for future statistical analysis data, similar to that called for in Problems 1 through 6, from sports, hobbies, stock market reports, individual research, etc.

3

PROBABILITY: SOME BASIC RESULTS

10 Probabilities are computed in many different ways depending on the circumstances under which they arise. Actually, the practicing statistician rarely has to compute probabilities on his own. Nearly all the probabilities that a statistician needs in daily work have already been computed and can be read from appropriate tables. Even so, for a better understanding every statistician should be familiar with certain basic principles and results.

In Chapter 2 we saw the close relationship between relative frequencies and probabilities. Since relative frequencies take values between 0 and 1, the same is true of probabilities. If A is any event, the probability of the event A has a value between 0 and 1, in symbols,

$$0 \leq P(A) \leq 1.$$

An event that never occurs has probability 0. An event that always occurs has probability 1.

11 Consider a student club at XYZ College whose membership is described in Table 3.1. The same information is presented graphically in Figure 3.2, where each small square represents one club member.

TABLE 3.1 **Membership of Student Club**

	Male	*Female*	*All Students*
Freshmen	20	20	40
Sophomores	18	12	30
Juniors	14	6	20
Seniors	8	2	10
All Classes	60	40	100

FIGURE 3.2 **Club Membership**

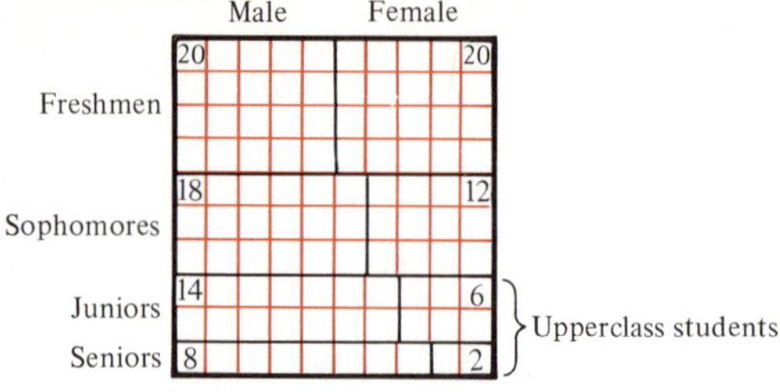

Suppose that the club has been asked to send one delegate to a conference on student affairs, and club members decide to select their representative by lot. Each member's name is to be written on a separate slip of paper. The 100 slips are to be mixed thoroughly in a large bowl, and the club president is to draw one slip from the bowl. We want to determine probabilities associated with events like "the student whose name is selected is female."

The experiment of selecting one name from among 100 names reminds us of an earlier experiment, the selection of one pingpong ball from among 10 (representing the digits 0, 1, . . . , 9). In that example we associated probability $\frac{1}{10}$ with any particular digit. Since we have 100 slips of paper in the present case, we associate probability $\frac{1}{100}$ with any particular slip. More generally, if the symbol F stands for the event that the name selected by our draw is that of a female student, we set $P(F) = \frac{40}{100} = .40$, since there are 40 females among the 100 members, and each has one chance in a hundred of having her name selected.

It is instructive to give this and similar probabilities a graphical interpretation. For this purpose, we shade the 40 squares that represent women members in Figure 3.3. If we express the shaded area as a proportion of the total area (the area enclosed by heavy lines), we find that this porportion equals $\frac{40}{100} = .40$, which is $P(F)$. By making the large square our unit of measurement, we can even say that $P(F)$ is equal to the area representing females. In future discussions we will often find it helpful to express probabilities as *areas*.

Right now let us look at two more probabilities associated with the selection of a member of our student club. (1) What is the probability that the selected person is an upperclass student (male or female)? (2) What is the probability that the selected person is a male senior? From Table 3.1 or Figure 3.2, we find that the first probability equals $\frac{30}{100} = .30$, since there are 30 upperclass students. Similarly, the second probability equals $\frac{8}{100} = .08$. However, these straightforward solutions hide two important aspects of our problems. To see what they are, let us introducc thc following symbols in addition to F:

FIGURE 3.3

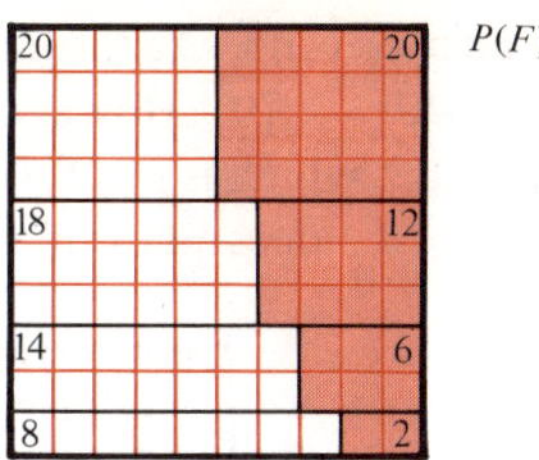

M for male
J for junior
S for senior
U for upperclass student
V for male senior.

In terms of these symbols, we can then replace the event U by the event "J or S," since an upperclass student must either be a junior *or* a senior. Similarly, we can replace the event V by the event "M and S," since a male senior must be both male *and* a senior. The problem of probabilistic interest is the following: Can we express $P(J \text{ or } S)$, which equals $P(U)$, in terms of $P(J)$ and $P(S)$? And can we express $P(M \text{ and } S)$, which equals $P(V)$, in terms of $P(M)$ and $P(S)$? We investigate the first question in Section 12 and the second question in Section 14.

The Addition Theorem of Probability

12 The question of whether $P(U) = P(J \text{ or } S)$ can be expressed in terms of the individual probabilities $P(J)$ and $P(S)$ is answered by a look at Figure 3.4, in which the squares representing upperclass students have been shaded. As we have seen, areas represent probabilities, and this particular area is the sum of two areas corresponding to juniors and seniors, respectively. Thus $P(J \text{ or } S) = P(J) + P(S)$. This result can be verified by direct computations based on Table 3.1. Indeed, $P(J) = .20$ and $P(S) = .10$, so that $P(J) + P(S) = .30 = P(J \text{ or } S)$.

The next question we should ask ourselves is whether the same simple relationship holds for any two events. The following example shows

FIGURE 3.4

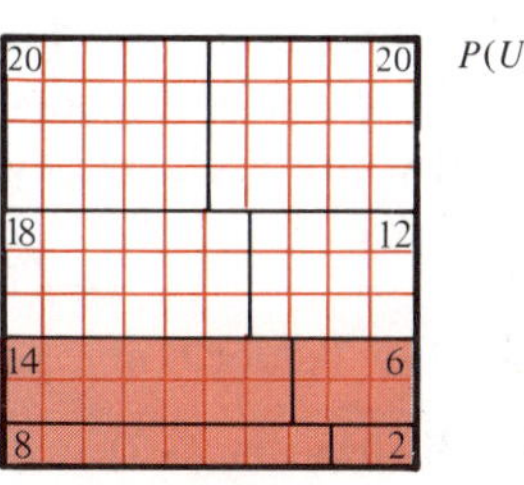

that the answer is no. According to earlier computations, $P(U) + P(F) = .30 + .40 = .70$. However, according to Figure 3.5, $P(U \text{ or } F) = {}^{62}\!/_{100} = .62$, which is less than .70. Thus $P(J \text{ or } S)$ equals $P(J) + P(S)$, while $P(U \text{ or } F)$ does not equal $P(U) + P(F)$.

FIGURE 3.5

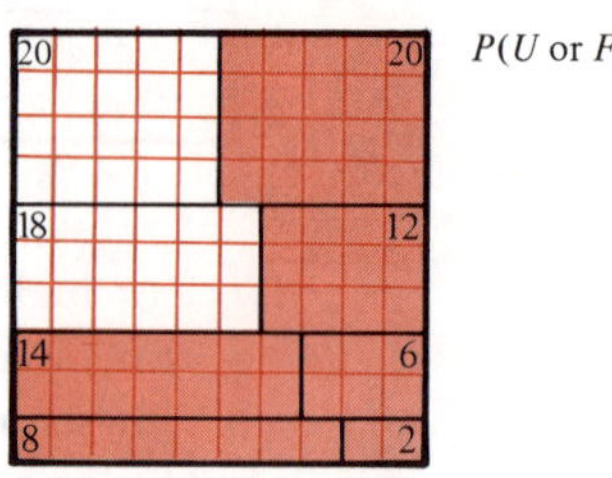

The reason for these two different results is not difficult to discover. Events J and S are **exclusive:** A student who is a junior cannot at the same time be a senior. Events U and F are not exclusive: A female student may well be an upperclass student or vice versa. Graphically, as Figure 3.4 shows, the areas representing juniors and seniors do not overlap, while according to Figures 3.3 and 3.4 the areas representing upperclass students and female students do overlap.

We have the following general result, which is the *Addition Theorem:* If events A and B are exclusive, then

3.6 $$P(A \text{ or } B) = P(A) + P(B).$$

For any two events A and B,

3.7 $$P(A \text{ or } B) = P(A) + P(B) - P(A \text{ and } B),$$

as we see in Problem 18 of this chapter.

Let us use (3.6) to derive an important result. Let A be any event and A^* denote the event "not A," so that A and A^* are *complementary* events. Clearly, $P(A \text{ or } A^*) = 1$, since either A or A^* is bound to occur. Also, since A and A^* are exclusive, $P(A \text{ or } A^*) = P(A) + P(A^*)$, so that

$$P(A^*) = 1 - P(A).$$

We will frequently use this rather self-evident result. As an example, using M for male and F for female, we find that $P(M) = 1 - P(F)$.

Formula (3.6) can be extended to more than two events. For example, if A, B, and C are exclusive events (no two can occur at one and the same time), then

3.8 $$P(A \text{ or } B \text{ or } C) = P(A) + P(B) + P(C).$$

The generalization of (3.7) is complicated. We observe that if events

A, B and C are not exclusive, then

3.9 $$P(A \text{ or } B \text{ or } C) < P(A) + P(B) + P(C).$$

Conditional Probabilities

13 So far when computing probabilities like $P(U)$ and $P(M)$, we assumed that selection was to be made among all 100 members of the club. It is sometimes helpful to refer to probabilities computed in this way as *unconditional,* since no restriction is imposed on the possible choice of a delegate. However, such an assumption may not always be appropriate. Suppose that delegates are restricted to upperclass students. If again a delegate is to be selected by lot, only 30 slips of paper bearing the names of the 30 upperclass club members should be used. The probability of selecting a female delegate now is $^8/_{30} = .267$ as follows from Figure 3.10. It is customary to refer to this new probability as

FIGURE 3.10

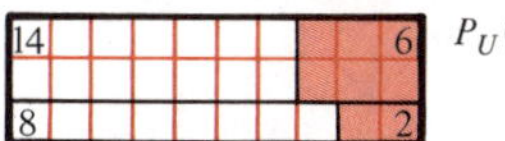

a conditional probability, since it has been computed subject to a supplementary condition, namely, that the delegate be an upperclass student. Notationally, we indicate the conditional character of the probability by adding an appropriate subscript that describes the nature of the condition. In the present case we write $P_U(F)$, which is read as "the probability of the event F computed under condition U" or, more briefly, "the probability of F, given U."

14 We turn now to the problem of expressing $P(S \text{ and } M)$ in terms of probabilities relating to the individual events, S and M. We remember that $P(S \text{ and } M) = {}^8/_{100}$, which we may write as $^8/_{100} = (^{10}/_{100})(^8/_{10})$. According to Figures 3.11 and 3.12, $^{10}/_{100} = P(S)$, the (unconditional)

FIGURE 3.11

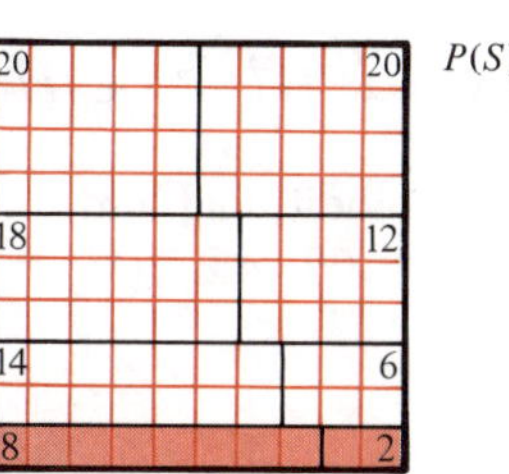

FIGURE 3.12 8 2 $P_S(M)$

probability of S, and $^8/_{10} = P_S(M)$, the conditional probability of M, given S. Thus

3.13 $$P(S \text{ and } M) = P(S)P_S(M).$$

What (3.13) says and Figures 3.11 and 3.12 show essentially is the following. The probability of the joint occurrence of events S and M can be computed in two stages. We first ask for the probability that the delegate be a senior; this probability is $P(S) = {}^{10}/_{100}$. Then having eliminated nonseniors from competition, we concentrate on the second requirement that the delegate be male. If we choose among senior club members, the probability $P_S(M)$ of selecting a male delegate is $^8/_{10}$.

We have the following general result, which is the *Multiplication Theorem:* For two arbitrary events A and B,

3.14 $$P(A \text{ and } B) = P(A)P_A(B)$$

In (3.14) the role of events A and B can be reversed to give

3.15 $$P(A \text{ and } B) = P(B \text{ and } A) = P(B)P_B(A).$$

Thus $P(S$ and $M)$ can also be computed as

$$P(M)P_M(S) = \frac{60}{100} \times \frac{8}{60} = .08$$

as before. Figure 3.16 provides the details for the computation of $P_M(S)$.

FIGURE 3.16

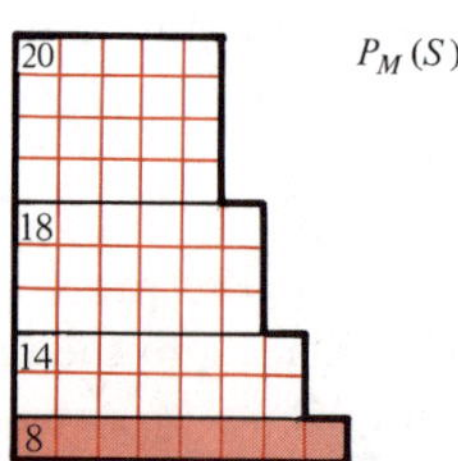

Independent Events

15 In general, conditional and unconditional probabilities differ in value. Thus $P(M) = .60$ and $P_S(M) = .80$. Assume that for two events, A and B, we have

$$P_A(B) = P(B),$$

that is, the conditional probability of B, given A, equals the unconditional probability of B. We then must also have

$$P_B(A) = P(A),$$

and we say that events A and B are independent. For independent events, the occurrence (or nonoccurrence) of either event does not change the probability of the occurrence (or nonoccurrence) of the other event.

If events A and B are independent, the Multiplication Theorem is particularly simple. From (3.14) and (3.15) we obtain the following result: If events A and B are independent, then

3.17 $$P(A \text{ and } B) = P(A)P(B).$$

Formula (3.17) can also be taken as definition of the independence of events A and B.

EXAMPLE 3.18 Suppose that we select one card from a well-shuffled deck of playing cards. (A regular deck containing 52 cards in 4 suits and 13 denominations is assumed.) Let Q be the event that the selected card is a queen and R the event that the card belongs to a red suit. As in other examples, we find the following probabilities,

$$P(Q) = \frac{4}{52} = \frac{1}{13}$$

$$P(R) = \frac{26}{52} = \frac{1}{2}$$

$$P_R(Q) = \frac{2}{26} = \frac{1}{13} = P(Q)$$

$$P_Q(R) = \frac{2}{4} = \frac{1}{2} = P(R)$$

$$P(\text{red queen}) = P(Q \text{ and } R) = \frac{2}{52} = \frac{1}{26} = P(Q)P(R).$$

Any one of the last three statements implies that events Q and R are independent.

16*

Independent events occupy an important place in probability and statistics. If all probabilities associated with an experiment are completely known, we can determine if two events A and B are independent, as

*An asterisk beside certain section numbers throughout the book indicates a section included only for enrichment. In the remainder of the book, no specific reference is made to any of the material discussed in the starred sections except possibly in some of the starred problems.

in Example 3.18. In statistical applications the situation may differ. We may need to know the probability $P(A \text{ and } B)$ associated with events A and B. If experimental conditions are such that it seems appropriate to assume that events A and B are independent, (3.17) determines the value of $P(A \text{ and } B)$. Thus it is important to know under what conditions it is realistic to assume that two events are independent. Let us see what the frequency interpretation of probability suggests.

If E is an event, we write $\#(E)$ to denote the number of occurrences of E in n trials. According to the frequency interpretation, $P(E)$ approximately equals the relative frequency of event E in a large number of trials. We can write this as

$$P(E) \doteq \frac{\#(E)}{n},$$

where $\doteq$ denotes approximate equality.

By definition, events A and B are independent if

$$P_A(B) = P(B).$$

If we repeat our experiment n times, and n is large, we have approximately

$$P(B) \doteq \frac{\#(B)}{n}.$$

Correspondingly, since $P_A(B)$ refers to experimental conditions where it is known that event A has occurred, we have

$$P_A(B) \doteq \frac{\#(A \text{ and } B)}{\#(A)}.$$

If then

3.19 $$\frac{\#(A \text{ and } B)}{\#(A)} \doteq \frac{\#(B)}{n},$$

we may assume that $P_A(B)$ equals $P(B)$. (3.19) says that the relative frequency with which B occurs among the trials that produce A is approximately equal to the relative frequency of occurrences of B among all n trials. The fact that A has or has not occurred does not change the frequency of the occurrence of B.

EXAMPLE 3.18 *cont'd.* Consider again the setup of Example 3.18 with R playing the role of A and Q playing the role of B. Then $\#(Q)/n$ equals the relative frequency of observing a queen in n draws of a single card from a regular deck of cards and $\#(R \text{ and } Q)/\#(R)$ equals the relative frequency of observing a queen in those of the n draws that produce a red card. Since red and black cards occur symmetrically in a regular deck of cards, the two relative frequencies may be expected to be approximately equal, sug-

suggesting independence of events Q and R. In the first part of Example 3.18, we were able to give a formal proof of the independence of the two events. In practical applications, sufficient information for a formal proof is rarely available.

17* *Total Probability*

Many probability computations require the combined application of both the addition and multiplication theorems. Let us return once more to the student club in Section 11 of this chapter. The delegate to be selected by the club will have to vote for or against a certain proposition while attending the student affairs conference. A survey of club members reveals that three out of four male members, but only one out of two female members, favor the proposition. What is the probability that the club delegate will vote in favor of the proposition? Let us denote this probability by Π. According to the addition theorem (3.6), we have

$$\begin{aligned}\Pi &= P(\text{delegate favors proposition})\\ &= P(\text{delegate is male and favors proposition})\\ &\quad + P(\text{delegate is female and favors proposition})\end{aligned}$$

If we now apply the multiplication theorem (3.14), we get

$$\begin{aligned}\Pi &= P(\text{delegate is male})P(\text{delegate favors proposition})\\ &\quad + P(\text{delegate is female})P(\text{delegate favors proposition})\\ &= .6 \times .75 + .4 \times .50 = .45 + .20 = .65,\end{aligned}$$

since the probability of a male delegate is .6 and the probability of a female delegate is $1 - .6 = .4$. Further, male delegates favor the proposition with probability .75, while female delegates favor the proposition with probability .50.

It is useful to state the previous result in more general terms. We are given two "strata" A and B, which are mutually exclusive and exhaustive, and an event E that may occur as the result of the selection of either strata A or strata B. (In the student club example, A and B refer to male and female club members; E is the event that the selected delegate favors the proposition.) Let $P(A)$ be the probability of selecting strata A and $P(B) = 1 - P(A)$ be the probability of selecting strata B. Our previous argument shows that

3.20 $$\Pi = P(E) = P(A)P_A(E) + [1 - P(A)]P_B(E),$$

where $P_A(E)$ is the (conditional) probability of the event E in strata A and $P_B(E)$ is the (conditional) probability of the event E in strata B.

Formula (3.20) is sometimes called the *Theorem of Total Probability.* The following example further illustrates its applicability. Twins are of two kinds, identical and fraternal. Identical twins are always of the same sex. The sex of fraternal twins is determined independently of each other. We want to find the probability that two twins are of different sex. In this example strata A and B refer to whether the twins are identical or fraternal. It is known that approximately one-third of all human twins are identical twins. So we set $P(A) = 1/3$. Formula (3.20) then becomes

$$\Pi = P(\text{one boy and one girl}) = \frac{1}{3} \times 0 + \frac{2}{3} \times \frac{1}{2} = \frac{1}{3},$$

since for identical twins we have zero probability that the two twins are of different sex, while for fraternal twins this probability is $1/2$, the same as for any two children who are not identical twins. We then also have

$$P(\text{two boys}) = P(\text{two girls}) = \frac{1}{3},$$

since the sum of all three probabilities must equal 1.

Another Look at Random Numbers

18 We have seen that in a table of random numbers, all digits from 0 to 9 have the same probability, $1/10$, of appearing in a given position. But random digits can also be read in groups of 2 or groups of 3 and so on. We can then ask ourselves what the probability is that a sequence of two random digits forms a number like 53. We have

$$P(53) = P(\text{first digit} = 5 \text{ and second digit} = 3).$$

But what is the probability on the right? Our method of producing random digits certainly implies that successive digits are independent. The reason for mixing our pingpong balls very thoroughly before drawing out another ball is to ensure independence of successive draws. Thus $P(53) = P(5)P(3) = (1/10)(1/10) = 1/100$, and a corresponding result holds for groups of three or more random digits. As an example,

$$P(539) = P(5)P(3)P(9) = \frac{1}{1000}.$$

When looking at random-number tables, people are often surprised to find double or triple or even quadruple entries like 33 or 555 or 8888. Somehow our minds seem to feel that such sequences have no place in a random scheme. But let us look at the question from the probabilistic point of view.

We have seen that all two-digit random numbers, like 53 or 87 or 22 have the same probability, $1/100$. Since there are 10 double numbers, 00, 11, and so on, the probability of having a two-digit random number with two identical digits is $10/100 = 1/10$. Thus on the average one of every 10 pairs of random digits should be a double number. There is an even simpler and more instructive way of deriving this probability by remembering how we obtained random numbers with the help of pingpong balls. We are asking for the probability that on the second try we draw the same ball as on the first try. Since there are 10 balls from which to choose, the probability is $1/10$. A similar argument shows that the probability of identical triplets of digits is $1/100$, and the probability of identical quadruplets is $1/1000$.

19 Students in an elementary statistics course were asked to write down what they considered to be a sequence of random digits. More exactly, they were asked "to mentally draw a sequence of pingpong balls." To simplify matters, only three balls labeled 1, 2, and 3 were to be used in this mental game. The record of 900 two-digit sequences produced in this way is given in Table 3.21.

TABLE 3.21 **Frequency of Guessed-digit Pairs**

		Second Digit 1	2	3	*Totals*
First Digit	1	62	144	119	325
	2	114	54	127	295
	3	106	116	58	280
Totals		282	314	304	900

We should like to find out whether real and imagined draws of pingpong balls produce similar results. For real draws, the probability is $1/3$ of obtaining each of the digits 1, 2, and 3. Thus we would expect to find something like 300 1s, 2s, and 3s, each, in 900 draws. If we look at the column on the right, we see that the students participating in the experiment wrote down 325 1s, 295 2s, and 280 3s as their first digit. Most of us will agree that these results look perfectly reasonable. We will get statistical evidence to this effect in a later chapter. The same is true of the frequencies for the second digit given in the bottom row of our table.

But let us now look at pairs of numbers. There are nine possible pairs: 11, 12, 13, 21, 22, 23, 31, 32, and 33. For real draws, each pair has the same probability, $(1/3)(1/3) = 1/9$. Thus in an experiment involving real pingpong balls, we would expect to find about 100 observations in each one of the nine cells. When we look at the experimental data, it is fairly obvious that there are not enough entries down the diagonal from the upper left to the lower right and that there are too many entries off the diagonal. Evidently, students were hesitant to write down pairs of identical numbers in their random sequences.

For the sake of comparison, the results of an experiment using real pingpong balls is given in Table 3.22. The totals are very similar to those in Table 3.21, but the entries in the body of the table are completely different. This time there is no particular difference between the entries down the diagonal and those off the diagonal. This, of course, is as it should be for random pairs of digits.

TABLE 3.22 **Frequency of Random-digit Pairs**

		Second Digit			
		1	2	3	*Totals*
First Digit	1	94	89	118	301
	2	104	90	87	281
	3	108	107	103	318
Totals		306	286	308	900

We will return to this example on several later occasions after we have discussed appropriate statistical methods for analyzing this type of data. Such a statistical analysis will not cause us to change our present impressions. It will simply help to put them on firmer foundations.

The reason for discussing the example now is to show the difference between independence and dependence. Successive digits produced by real draws of pingpong balls are independent (provided sufficient mixing of balls between successive draws takes place). The same is not true of mental draws. Our minds remember earlier choices. This remembrance, though possibly unintentional, introduces an element of dependence.

We can draw an important conclusion from the experiment. Our minds do not seem to be able to imitate real random behavior. In Chapter 1 we said that statistical inference is concerned with generalizations based on sample information. However, statistical procedures provide a basis for valid generalizations only if the sample information has been obtained according to known laws of the theory of probability. It follows that, in collecting sample information, it is advisable to use mechanical means like random-number tables or similar devices to ensure proper random selection. Experiments like the one we have just discussed show that people cannot be trusted to obtain truly random samples if the actual selecting is left to them.

20 *Application to Sampling*

In conclusion then, we want to indicate how random numbers can be used to obtain random samples. Consider the following rather common problem. We want to survey student attitude at a given university on some topic such as the legalization of marijuana. By one method or another, we decide to include 225 students in our sample. How are we going to select them? We might simply take the first 225 students entering the dining room for breakfast one morning. But what about the students who eat at home or do not feel like eating breakfast that morning? It is quite possible that students living at home feel differently about smoking marijuana than students living in dormitories.

We want a method of selection that chooses students one at a time and with equal probability. The problem is quite simple if there exists a central file of all students at the university. We can then assign each student an identification number. Indeed, at many universities students already have identification numbers. For simplicity, let us assume that these numbers go consecutively from 1 to, say, 7629. Since the highest possible number has four digits, we take a table of random digits and consider them in blocks of four. Suppose that this is the way our table looks:

. . . . . 18048 25400 76364 . . . . .

The first student to be included in our sample is the one with identification number 1804. Since no student has the number 8254, we skip this number. The next student has number 0076, or simply 76. We continue in this way until we have obtained 225 identification numbers. Of course, now the real work begins. We have to get the names and addresses of the 225 students and contact them. Other methods of selection may be simpler and faster, but this method has the all-important property of true randomness.

21* When using a random-number table to select a random sample, it may happen that the same number appears more than once. Two possibilities are then open to us. We can record a student's opinion as often as that student's identification number appears in the table of random numbers. The resulting sampling procedure is known as sampling with replacement. Alternatively, we can decide to ignore a number after its first appearance exactly as we ignore a number that does not correspond to a member of our student population. The resulting sampling procedure is known as sampling without replacement. Before taking a sample, we have to decide which of the two sampling procedures is appropriate under the circumstances.

EXAMPLE 3.23 A new student organization has ten members, seven men and three women. They decide to leave the election of a president and secretary to chance. What is the probability that both offices will be filled by women?

If the students do not mind having both offices filled by one and the same person, they can use sampling with replacement. Under sampling with replacement successive choices are independent, and the desired probability is simply

$$\frac{3}{10} \times \frac{3}{10} = \frac{9}{100} = .090.$$

If the students want to be sure that the offices of president and secretary are filled by two different persons, sampling with replacement is inappropriate. Under sampling without replacement, computations are slightly more complicated. We now need the formula $P(A \text{ and } B) = P(A)P_A(B)$ for the simultaneous occurrence of two events that are not independent. In the present example A is the event that the student elected to the office of president is a woman, while B is the event that the student elected to the office of secretary is a woman. We have $P(A) = 3/10$ and $P_A(B) = 2/9$, so that

$$P(\text{both offices are filled by women}) = \frac{3}{10} \times \frac{2}{9} = \frac{6}{90} = .067$$

when sampling without replacement.

The Taxi Problem Revisited

22 In Chapter 1 we mentioned briefly the need for setting up an appropriate framework—or *probability model*, as we will say from now on—for solving a statistical problem. Two assumptions we mentioned in the taxi problem, consecutive numbering starting with 1 and equal likelihood for any one taxi to pass the spot where we are standing, may be translated into mathematical language by saying that the five taxi numbers in Chapter 1 constitute a random sample of size 5 from a population of tags numbered from 1 to 550. As we have seen, such a sample can be obtained by selecting 5 three-digit numbers from a random-number table ignoring combinations 551–999 and 000. Indeed, the samples in Chapter 1 were obtained in exactly this fashion.

We want to investigate more precisely why, in Chapter 1, we decided to reject the hypothesis that there are at least 1000 taxis in the given city. For our investigation, we assume that the random-number model is appropriate for analyzing taxi data. Let us assume for the moment that there are exactly 1000 taxis. In our mathematical model these taxis are represented by the three-digit numbers 001, 002, . . . , 999, 000 (in place of 1000). The largest observed taxi number in our sample was 440. Let us then compute the probability that each of five three-digit random numbers selected from all three-digit numbers is 440 or smaller. The probability that any one such number is smaller than or equal to 440 is $440/1000 = .44$. Our model implies independence from one selection to another. Therefore, the probability that all five numbers are smaller than or equal to 440 is $(.44)^5 = .016$. Thus according to the frequency interpretation of probability, we should expect to observe such an event only about 1.6 times in 100 trials if, indeed, there are 1000 taxis in the city, and even less frequently if there are more than 1000 taxis. It is always possible that we actually did observe an event having probability .016 or less. On the other hand, a much more plausible explanation is that the number 1000 is incorrect. As a consequence, we decide that the claim that there are at least 1000 taxis at our disposal is very doubtful. In statistical terminology, we decide to *reject* the hypothesis being tested.

23 It is instructive to pursue the problem somewhat further. We have already rejected the hypothesis that there are at least 1000 taxis. But what about claims of at least 900, or 800, or perhaps only 700? Corresponding hypotheses can be tested in very much the same way. Let us denote the correct number of taxis by t. If $t = 900$, the probability that any one taxi number is smaller than or equal to 440 is $^{440}/_{900} = .49$, and the probability that all five taxi numbers are smaller than or equal to 440 is $(.49)^5 = .028$, still rather small. Very likely, there are not even 900 taxis around. When we come to $t = 800$, we find that $^{440}/_{800} = .55$ and $(.55)^5 = .050 = {}^1/_{20}$. As we will see in later chapters, statisticians often use $^1/_{20}$ as a dividing point between "small" probabilities, suggesting rejection of a hypothesis, and sufficiently large probabilities that have no such implication. If we follow this practice, we would then claim that t is smaller than or equal to 800, or in symbols, $t \leq 800$. We have then found a "reasonable" upper bound for the number of taxis in the city.

What about a lower bound? Assuming the same probability model as before, we are certain that t is at least equal to 440, since we actually observed taxi #440. On the other hand, if $t = 440$, the probability that we would observe taxi #440 in a sample of five is only .011 (see Problem 14 of this chapter). So presumably t is larger than 440. Using the kind of argument that led to the upper bound 800 and computations that are only slightly more complicated, we find that a reasonable lower bound for t is 444, in symbols, $444 \leq t$ (see Problem 15 of this chapter). By combining our two statements, on the basis of what we have observed we can be reasonably confident that the true number of taxis is somewhere between 444 and 800,

$$444 \leq t \leq 800.$$

Such a statement is called a confidence interval for the quantity t. Confidence intervals are also referred to as interval estimates in contrast to the point estimates discussed in Chapter 1. When we estimated t

as 512 in Chapter 1, we realized of course that this estimate deviated more or less from the true value t. There we were in the fortunate position of being able to find out the true value of t. But, in general, that possibility does not exist. In such a case the point estimate does not provide any indication of the magnitude of its possible error. On the other hand, a confidence interval has a built-in indicator of its accuracy, or possibly inaccuracy—namely, the length of the interval. A relatively long interval provides very little information; a short interval provides a good deal of information. In our particular case the interval extends from 444 to 800 showing that a sample of 5 provides only very limited information about t.

PROBLEMS

1 A college has 1000 students who have indicated the following preferences for their area of concentration, where N denotes a concentration in the natural sciences, S, in the social sciences, and H, in the humanities:

		Area of Concentration			
		N	*S*	*H*	*Totals*
Class	*Fr*	75	125	100	300
	So	60	100	90	250
	Ju	50	110	90	250
	Se	45	85	70	200
	Totals	230	420	350	1000

In addition to the indicated symbols, set $U = (Ju \text{ or } Se)$, $L = (Fr \text{ or } So)$. Find the following probabilities associated with the random selection of a student. (Always express verbally the event whose probability you are determining.)

a. $P(N)$, $P(S)$, $P(H)$;
b. $P(Fr \text{ and } N)$, $P(Ju \text{ and } S)$, $P(Se \text{ and } H)$;
c. $P(L \text{ and } H)$, $P(U \text{ and } N)$;
d. $P(N \text{ or } S)$, $P(U \text{ or } H)$;
e. $P_{Se}(N)$, $P_N(Se)$, $P_{Ju}(N \text{ or } S)$, $P_U(H)$.

2 Use the same information as for Problem 1 of this chapter.

a. Name several pairs of events that are exclusive.
b. Name two events that are independent.

3 An insurance company analyzed hospital use of 1000 men and 1000 women insured under a hospital plan. These are the findings:

	Men	*Women*
Used Hospital	100	150
Did Not Use Hospital	900	850

a. What is the probability that an insured person uses the hospital?
b. Is hospital use independent of the sex of the insured?

4 Consider the following events associated with the example discussed in Section 11 of this chapter: *M, F, U, J, S, J* or *S, J* and *M, J* and *F, S* and *F*. Find three events that are exclusive. (No two can occur at the same time.)

5 We have 900 slips of paper corresponding to the 900 guesses tabulated in Table 3.21. One of the slips is selected at random.

a. What is the probability that the second digit is a "1"?
b. What is the probability that the two digits are equal?
c. If the first digit is a "1," what is the probability that the second digit is a "3"?

6 Two independent events A and B have probabilities $P(A) = 1/3$ and $P(B) = 2/3$. Find

a. $P(A \text{ and } B)$;
b. $P_B(A)$.

7 Events C and D are exclusive with $P(C) = P(D) = 1/3$. Find

a. $P(C \text{ or } D)$;
b. $P(C \text{ and } D)$;
c. $P_D(C)$.

8 E and F are two events with probabilities $P(E) = 2/5$ and $P(F) = 3/5$. Can you find

a. $P(E \text{ and } F)$?
b. $P(E \text{ or } F)$?

9 A football coach complains that three games in a row his team has lost the toss at the start of the game. Does the coach have reason to complain?

10 A newspaper regularly publishes the winning lottery numbers for the state lotteries of Connecticut, Massachusetts, New Jersey, New York, and Pennsylvania. In one particular week all five numbers start with the same digit. What is the probability of this happening due to chance?

11 What is the probability that among five random numbers, all known to lie between 1 and 60 (both limits included), all are greater than 20?

12 Consider the first sample of taxi numbers in Table 1.1. Does 660 seem a reasonable value for the total number t of taxis? Why?

13 Use a table of random digits, such as Table J, to find five taxi number samples of size 3 each, assuming that there are 75 taxis in all. For each of your samples, compute the extreme and median estimates. Which estimate would you say is closer to the true value?

14 There are 440 taxis in a city, numbered from 1 to 440. A random sample of five taxis is observed. Show that the probability that taxi #440 appears at least once in the sample is .011. (*Hint:* Find the probability of the complementary event, that is, the event that #440 does not appear in the sample.)

15 There are 444 taxis in a city, numbered from 1 to 444.

a. Show that the probability is .055 that, in a random sample of five taxis, the largest observed taxi number is 440 or greater. (*Hint:* See Problem #14 of this chapter.)
b. What is the probability of this event if there are only 443 taxis in the city?

16 How can a table of random numbers be used to simulate the pingpong ball experiment that produced the data for Table 3.22?

17 Consider the experimental setup of Example 3.18, but assume that one card is missing from the deck. Show that the events Q and R are not independent assuming that

a. the missing card is red, but not a red queen
b. the missing card is neither a red nor a black queen.

18*Let events U and F have the same meaning as in Section 11. With the help of Figures 3.3, 3.4, and 3.5, show that the following statement holds:

$$P(U \text{ or } F) = P(U) + P(F) - P(U \text{ and } F).$$

Verify the statement by substituting numerical values for the four probabilities.

The corresponding result for any events A and B is Formula (3.7). Show that if events A and B are exclusive, (3.7) reduces to (3.6).

19*In a lottery you win a consolation prize if the first three or the last three digits of your five-digit number match the corresponding digits of the winning number. Find the probability that a person who has bought one lottery ticket wins a consolation prize. [*Hint:* Use Formula (3.7).]

20*Consider two-digit random numbers. Let G be the event that the number is divisible by 4; H, the event that it is divisible by 5.

a. Find $P(G)$ and $P(H)$.
b. Find $P(G \text{ and } H)$.
c. Are events G and H exclusive?
d. Are events G and H independent?
e. Find $P(G \text{ or } H)$. [*Hint:* Use Formula (3.7).]

21*Mr. X can drive home from work using either route A or route B. On route A he has one chance in five of being delayed by a traffic tie-up; on route B he has one chance in three. If Mr. X chooses route A with probability .8 and route B with probability .2, what is the probability that he will be delayed by a traffic tie-up?

22*When Ms. Y has a headache, she selects at random one of two headache remedies. If one remedy relieves three out of four headaches and the other relieves two out of three headaches, what is the probability that Ms. Y's headache is not relieved?

23*Use Formula (3.20) to prove directly that among twins $P(\text{two boys}) = P(\text{two girls}) = \frac{1}{3}$. [*Hint:* See the discussion in the example following Formula (3.20).]

4

BINOMIAL PROBABILITIES

24 In this and some of the following chapters, we consider experiments consisting of a sequence of trials at each of which a certain event does or does not occur. A single random digit is or is not greater than 4. A newly purchased car does or does not require major corrective work. A person selected for a survey is or is not a registered voter. Since each trial has only two possible outcomes, we speak of *binary* or *binomial* experiments. In this chapter we will discuss a simple probabilistic model that is often adequate for the statistical analysis of data resulting from binomial experiments. In Chapters 6–8 we will take up the actual statistical analysis.

Almost universally in mathematical and statistical literature, the two possible outcomes of a trial are given the names *success* and *failure.* In everyday life these two words have very specific meanings. In probability and statistics they are simply convenient names without the usual connotations. It is quite possible that the death of a patient is called a "success" if the investigator should be studying the fatality rate associated with a certain kind of operation. We denote the number of trials to be performed in a given experiment by n, and it is always understood that the value of n has been fixed before the start of the experiment. Thus the theory that we are going to develop does not apply if n is determined in the course of the experiment, as in the case of the gambler who looks at past winnings or losses and decides to withdraw from the game. We also always assume that successive trials are independent of each other—that is, success or failure on any one trial has no bearing on success or failure on any subsequent trials. Finally, we assume that the probability of success remains the same throughout the experiment. There are many experiments where one or both of these assumptions are violated. In such cases the theory that we are going to develop, if applied, may give quite erroneous answers.

We always use the letter p for the probability of success in a single trial; for the probability of failure, we write q. Since a trial has only two possible outcomes, success and failure, we must have $q = 1 - p$. Thus we have the following notation:

n = number of trials
p = probability of success in any one trial
$q = 1 - p$ = probability of failure in any one trial.

Furthermore, we can summarize our basic assumptions as follows.

(i) The number n of trials is fixed by the experimenter before the start of the experiment.
(ii) Successive trials are independent.
(iii) The success probability p does not change from one trial to the next.

Binomial Probabilities

25 Now we can state the problem we want to solve. If k is any integer between 0 and n, what is the probability, say $b(k)$, that, in an experiment consisting of n trials, k of the trials are successful, and $n - k$ are unsuccessful? Before we tackle this general problem, let us first look at a special case. At a country fair we win a prize if we roll either a 5 or a 6 with a single die. If we decide to try our luck three times, what is the probability that we will win exactly two prizes? If the die we are using is fair, we have two chances in six of winning a prize every time we roll the die. Let us agree to call winning a prize a success and not winning a prize a failure. We then want to solve the following problem: What is the probability of observing two successes and one failure in three trials with success probability $1/3$ at each trial?

It is a good idea to start by making a list of all the possible ways in which we can have two successes. The best way of doing this is to write down what happens at each of the three trials, letting S represent success and F, failure. This list is quite simple.

$$SSF \qquad SFS \qquad FSS$$

Our next problem is to find the probability of each of these three possibilities. Once we have done this, the final step is simple. Since the three possibilities are exclusive, according to the addition theorem (3.6) in Section 12, we add all three probabilities.

We take the case SSF first. Since we assume that successive trials are independent, we have, extending the multiplication theorem (3.17) for independent events in Section 15,

$$P(SSF) = P(S)P(S)P(F) = \frac{1}{3} \times \frac{1}{3} \times \frac{2}{3} = \frac{2}{27}.$$

Similarly, $P(SFA) = P(S)P(F)P(S) = \frac{1}{3} \times \frac{2}{3} \times \frac{1}{3} = \frac{2}{27}$,

and $P(FSS) = P(F)P(S)P(S) = \frac{2}{3} \times \frac{1}{3} \times \frac{1}{3} = \frac{2}{27}$.

The sum of these three probabilities is $6/27 = 2/9$, and this is the probability of winning exactly two prizes in our three attempts.

In exactly the same fashion we could compute, say, the probability of winning one prize in three trials. But let us do something more general. We have computed our probability under the specific assumption that $p = 1/3$—that is, that the die we are using is fair. But suppose that we want to know our chances if p is equal to .30 or .28, rather than .33. (Not all dice are necessarily fair.) Clearly, our method works for any specific value of p, not only for $p = 1/3$. Indeed, there is no reason at all why we cannot use the letter p instead of a numerical probability. We can actually use part of our earlier results:

$$P(SSF) = p \times p \times q = p^2q;$$
$$P(SFS) = p \times q \times p = p^2q;$$
$$P(FSS) = q \times p \times p = p^2q.$$

All three arrangements have the same probability p^2q. The reason for this is quite obvious. Each arrangement contains the letter S twice and the letter F once. And since, according to our assumption, successive trials are independent of each other, each S contributes p, and each F contributes q to the probability of the given arrangement. Thus the probability we are looking for is equal to the probability of any one of the possible arrangements multiplied by the number of possible arrangements, in our case, 3:

$$P(2 \text{ prizes}) = 3p^2q.$$

If we set $p = 1/3$ and $q = 2/3$, we should get our earlier result. Indeed, $3(1/3)^2\,(2/3) = 2/9$.

Let us now take the same approach for the case of 1 win and 2 losses. A typical arrangement giving this result is SFF and

$$P(SFF) = P(S)P(F)P(F) = pqq = pq^2.$$

There are two other arrangements—namely, FSF and FFS—that have the same probability. Thus

$$P(1 \text{ prize}) = 3pq^2.$$

To complete the picture,

$$P(3 \text{ prizes}) = P(SSS) = P(S)P(S)P(S) = ppp = p^3,$$
$$P(\text{no prize}) = P(FFF) = q^3.$$

Here we have brought together our results:

Number of Successes k	Probability $b(k)$ of k Successes	Special Case: $p = 1/3$
0	q^3	$8/27$
1	$3pq^2$	$4/9$
2	$3p^2q$	$2/9$
3	p^3	$1/27$

Since these probabilities refer to binomial experiments, they are called binomial probabilities. The tabulated probabilities form the *binomial distribution* with $n = 3$ and success probability p.

Let us use our results to solve the following problem. What is the probability that we win at least one prize in three trials? At least one prize means one, or two, or three prizes. These events are exclusive so that by the addition theorem (3.6) the desired probability equals $b(1) + b(2) + b(3)$. We can then substitute in this expression from our earlier table of probabilities. Actually, a more compact answer is possible. Since "at least one prize" and "no prize" are complementary events, we have

$$P(\text{at least one prize}) = 1 - P(\text{no prize}) = 1 - b(0) = 1 - q^3.$$

In particular, for an honest die we have

$$1 - \left(\frac{2}{3}\right)^3 = 1 - \frac{8}{27} = \frac{19}{27} \text{ or } .70.$$

A General Formula

26 We are now ready to solve the general problem. What is the probability, $b(k)$, that, in n trials with probability p of success in a single trial, exactly k of the trials end in success, and $n - k$ end in failure? As we have seen, we need only multiply the probability of any arrangement of k successes and $n - k$ failures by the number of possible arrangements. One such arrangement is $SS \cdots SFF \cdots F$, a sequence of k successes followed by a sequence of $n - k$ failures. This arrangement has probability

$$P(SS \cdots SFF \cdots F) = pp \cdots pqq \cdots q = p^k q^{n-k}.$$

It follows that

$$b(k) = C(n, k)p^k q^{n-k},$$

where the so-called *binomial coefficient* $C(n,k)$ tells us in how many different ways k letters S and $n - k$ letters F can be arranged.

The derivation of an appropriate formula for $C(n, k)$ is not difficult, but it is time consuming. Since many readers probably have seen a derivation before or will see one in another course, we will only provide the appropriate formula—namely

$$C(n, k) = \frac{n!}{k!(n - k)!},$$

where $n! = n(n - 1)(n - 2) \cdots (3)(2)(1)$ and $0! = 1$. For example, $5! = (5)(4)(3)(2)(1) = 120$. We then have

4.1 $$b(k) = \frac{n!}{k!(n - k)!} p^k q^{n-k}.$$

As a check, let us take $n = 3$, $k = 2$, and $p = 1/3$. Then

$$b(2) = \frac{3!}{2!1!}\left(\frac{1}{3}\right)^2\left(\frac{2}{3}\right) = \frac{2}{9},$$

which agrees with the result we obtained in our special problem involving the probability of winning two prizes in three trials with a fair die.

Theoretically, we can use our formula to compute the probability of k successes for any values of k, n, and p, but there are practical difficulties. The reader should try to compute probabilities for various values of k, n, and p to gain an appreciation of the amount of work that is required! Of course, with high-speed computers these computations are no problem. Whenever possible, statisticians use ready-made tables or programmed calculators. Since a statistician can never tell which values of n and p may be encountered, tables of the binomial distribution have to be very extensive.

Table A contains binomial probabilities for certain selected values of n and p. The tables included are for illustrative purposes only. For extensive statistical work, the student will have to consult special tables of the binomial distribution, such as the Harvard Tables cited in the Bibliography, or, where appropriate, use the normal approximation to be discussed in the next chapter.

EXAMPLE 4.2 What is the probability of observing four successes in ten trials with success probability .20? According to Formula 4.1,

$$b(4) = \frac{10!}{4!6!} .2^4 \times .8^6 = .0881.$$

More conveniently, the answer can be found to three-place accuracy in Table A as follows:

(i) go to the table for $n = 10$,
(ii) find the column labeled $p = .2$,
(iii) if—as in this case—p is listed in the *top* row, look for the appropriate k-value in the *left-hand* margin. The desired probability is listed at the intersection of the row labeled $k = 4$ and the column labeled $p = .2$: $b(4) = .088$.

EXAMPLE 4.3 What is the probability of observing 16 successes in 20 trials with success probability .70? We use Table A for $n = 20$. The value $p = .70$ is found in the *bottom* row. Therefore, we look for k in the *right-hand* margin. At the intersection of the row labeled $k = 16$ and the column labeled $p = .70$, we find $b(16) = .130$. (Note that the event "16 successes in 20 trials with success probability .70" can also be described as "4 failures in 20 trials with failure probability .30." Thus we could also have entered Table A with $n = 20$, $p = .30$, and $k = 4$. The way Table A has been arranged, the result is exactly the same.)

EXAMPLE 4.4 What is the probability of observing at least 16 successes in 20 trials with success probability .70? We have

$$\begin{aligned}
&P(\text{at least 16 successes}) \\
&= P(k \geq 16) \\
&= b(16) + b(17) + b(18) + b(19) + b(20) \\
&= .130 + .072 + .028 + .007 + .001 \\
&= .238.
\end{aligned}$$

PROBLEMS

1 Find $b(k)$ for $n = 3$, $p = 1/6$, $k = 0,1,2,3$.

2 Find $b(k)$ for $n = 4$, $p = 1/3$, $k = 0, 1, \ldots, 4$.

3 Find the probability $b(k)$, $k = 0, 1, \ldots, 5$, that, in five independent trials with success probability p, we observe k successes.

4 In ten trials with success probability .30, what is the most likely number of successes (the number having highest probability)?

5 If you roll a fair die three times, what is the probability that you will observe at least one "6"?

6 In ten binomial trials with success probability $1/3$, find the probability of

a. at least five successes;
b. at most two successes;
c. more than four successes;
d. fewer than five successes.

7 In ten binomial trials with success probability $2/3$, find the probability of

a. exactly six successes;
b. more than six successes;
c. fewer than six successes;
d. at least eight successes.

8 An examination consists of 25 questions, at least 17 of which have to be answered correctly for a passing grade. A student knows the answers to 60 percent of the type of questions likely to be on the examination. What is the probability of the student's receiving a passing grade?

9 A certain do-it-yourself job requires the use of 12 screws. It has been found that approximately 10 percent of all kits required for the job contain at least one faulty screw. A store sells 25 of the kits. Find the probability that the store will receive

a. no complaint about faulty screws;
b. at most two complaints.

What assumption(s) is (are) implied in your answer?

10 An automobile manufacturer announces that an estimated 30 percent of a certain line of cars have defective brakes requiring corrective action. A dealer handling the particular line of cars receives ten repair kits from the manufacturer. What is the probability that this dealer will be able to repair all defective cars, if 25 owners bring in their cars for inspection?

11 Two people sitting down at a square table may sit "opposite" like this x □ x, or "across corners" like this, □ x.
x

a. Show that, if two people choose at random among possible seating arrangements, the probability that they will sit opposite each other is $\frac{1}{3}$.
b. If ten couples sit down at ten tables, what is the probability that at least eight couples sit across corners (assuming random seating)?

12 A calculator salesperson claims that two out of three students prefer one brand of calculator to a competing brand. If the salesperson's claim is correct, what is the probability that, in a sample of 13 students, at least seven favor the competitor's model?

13 Coin tosses are said to be *fair* if the probability of the coin falling heads equals the probability of the coin falling tails. Show that any coin (even a badly worn coin) can be made to produce fair tosses as follows. The coin is tossed twice, and we observe whether it falls *TT*, *TH*, *HT*, or *HH*, where *T* stands for tails and *H* stands for heads. We ignore tosses resulting in *TT* or *HH*. If the coin falls *TH*, we say that "it has fallen heads"; if the coin falls *HT*, we say that "it has fallen tails."

14 A college football team has seven games on its schedule. What is the probability that the team will win the toss more often than not?

15 Assume that one tire in three of a certain brand of tire wears out before 20,000 miles. Use the binomial distribution to compute the probability that a person who uses four new tires of this particular brand will be able to use all four tires for at least 20,000 miles. Can you think of a reason why the binomial distribution may not give a very reliable answer in this case?

16 In the World Series the team that wins four out of seven games wins. Can you think of reasons why, in general, the binomial model is inappropriate to describe World Series play?

17 The passengers on an airplane can leave the plane through the front or the rear door. If n passengers want to get off the plane, do you think that the binomial distribution for n trials and success probability $\frac{1}{2}$ is appropriate to compute probabilities involving the numbers of passengers that are going to leave the plane through the rear door? Why?

5

THE NORMAL DISTRIBUTION

An Approximation for Binomial Probabilities

27 It is often said that one picture is worth a thousand words. To most mathematicians a mathematical formula is even better than a picture. However, there are times when pictures and graphs are very useful to interpret concretely what a formula has to say. For example, consider the formula for binomial probabilities that we discussed in Chapter 4,

$$b(k) = \frac{n!}{k!(n-k)!} p^k(1-p)^{n-k},$$

for $k = 0, 1, \ldots, n$. In particular, for $n = 3$ and $p = 1/3$, we can compute the successive probabilities

$$b(0) = \frac{8}{27};$$

$$b(1) = \frac{12}{27};$$

$$b(2) = \frac{6}{27};$$

$$b(3) = \frac{1}{27}.$$

A picture can present much more vividly how these probabilities change with k. The customary way to present such probabilities is by means of a *histogram*. We plot the possible values of k along a horizontal axis and then erect rectangles with base centered at k and height equal to $b(k)$ as in Figure 5.1. Since each rectangle has a base of length 1, the areas of the various rectangles are our probabilities $b(k)$. The eye

FIGURE 5.1

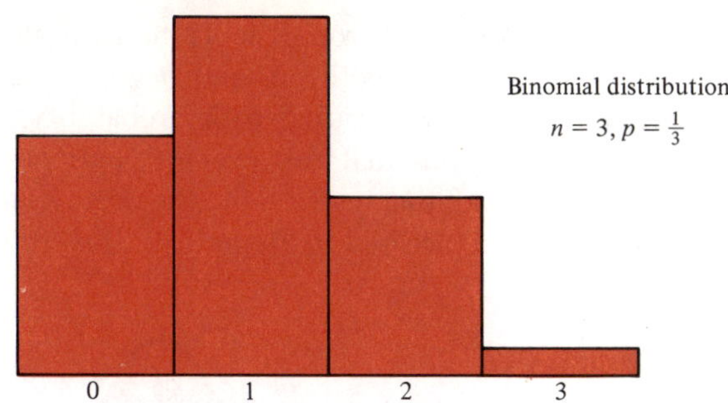

concentrates automatically on areas, and in a histogram we get a very graphic picture of the distribution—that is, the way in which the total probability 1 is distributed among the possible values of k.

It is interesting to study how a change in the number of trials affects the probabilities $b(k)$. In many courses on probability a great deal of time and effort is expended on such studies resulting in important theorems. In this course we are content to use the results of such efforts. Before we do so, we want to indicate graphically how these results come about. For this purpose, we look at two more histograms of binomial distributions again with success probability $\frac{1}{3}$ but with different values of n. In Figure 5.2 we have $n = 12$ and in Figure 5.3,

FIGURE 5.2

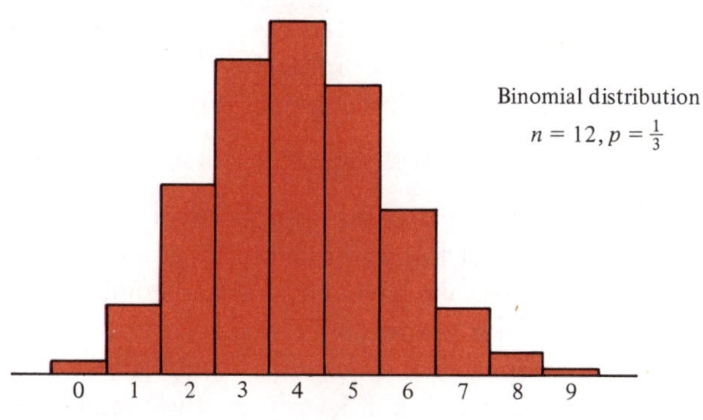

FIGURE 5.3

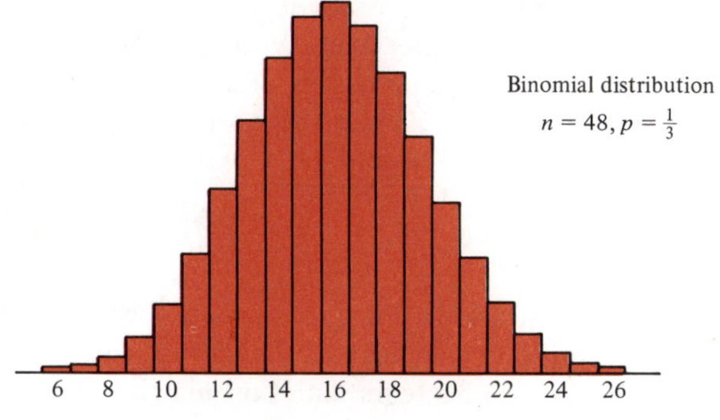

$n = 48$. As we go from 3 to 12 to 48 trials, we note a striking fact: The histograms smooth out and begin to take on a more symmetric shape. In courses dealing with probability, it is shown by advanced calculus methods that this process continues as the number of trials increases, and that eventually the histograms tend to the so-called *normal curve* pictured in Figure 5.4. If we trace the normal curve and

FIGURE 5.4

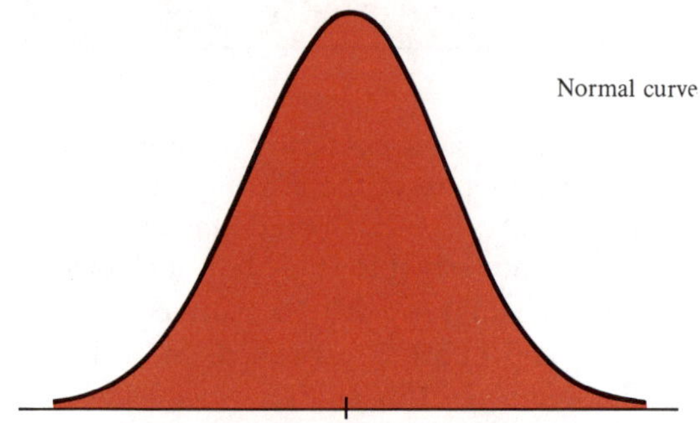

superimpose it on the histograms (centering it at the points 1, 4, and 16, respectively), we find that it gives a rather bad fit for $n = 3$, a considerably improved fit for $n = 12$, and a very satisfactory fit for $n = 48$.

Our reasons for studying these results are primarily practical. Suppose that we are interested in computing the probability that in 48 trials with success probability $1/3$ we observe between 10 and 25 successes, both limits included. From an appropriate table we find

$$P(10 \leq k \leq 25) = b(10) + \cdot \cdot \cdot + b(25) = .978.$$

But suppose that an appropriate table is not available. To compute each one of the 16 probabilities from our formula is a rather hopeless task, unless we have a high-speed computer at our disposal. What else can we do? It turns out that our new discovery provides us with a means of computing the desired probability, at least approximately. A look at Figures 5.3 and 5.4 reveals the method. We know that each probability $b(k)$ is represented by the area of a rectangle. This means that the probability we are looking for is represented by the area of a sum of rectangles, namely, those labeled 10, 11, and so on up to 25. Let us approximate this area by the corresponding area under the normal curve. Since the normal curve fits rather closely over the histogram, we should get a reasonably good approximation to the desired probability.

28 But how are we going to compute the area under the normal curve? In a course in calculus it is shown that areas under a curve are computed by means of integration. Fortunately, we do not have to compute an integral every time we want to use the normal curve, which we will frequently want to do. Instead, we refer to appropriate tables. Two such tables and their uses are described in subsequent sections, but first some general remarks are in order.

From our discussion it may appear that there is just one normal curve. This is not so. A normal curve depends on two constants which are called the *mean* and the *standard deviation,* customarily represented by the symbols μ (Greek mu) and σ (Greek sigma), respectively. The mean μ can be any number; the standard deviation σ can be any positive number.

The meaning of these two constants is best seen from Figure 5.5.

FIGURE 5.5

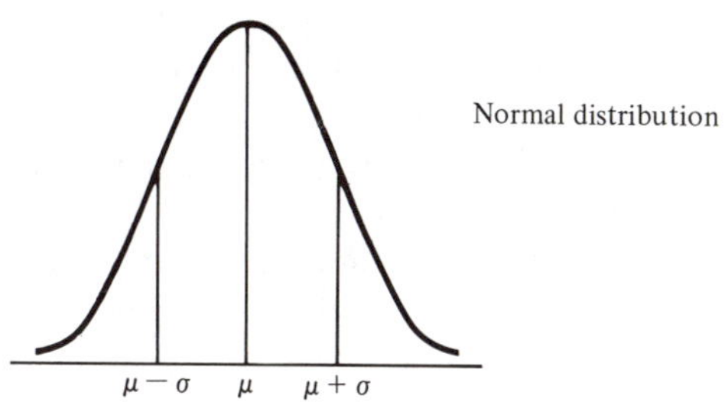

The mean μ tells us where the center of symmetry of the normal curve is located. The meaning of the standard deviation σ is more complicated. Graphically, σ is the distance from the center of symmetry to the two points at which the curve changes from curving down to curving up. A change in the value of μ does not change the outline of the curve; rather it moves the whole curve farther to the right or left depending on whether μ increases or decreases. A change in σ, on the other hand, changes the outline of the curve. The curve is always symmetric about μ, but for small values of σ, it has a high and narrow outline, and for large values of σ, its outline is low and flat.

We are going to use areas under a normal curve to compute probabilities. Since the curve tells us how probability is distributed over various intervals, we will prefer usually to speak of normal distributions rather than normal curves. More specifically, we will speak of the normal distribution with mean μ and standard deviation σ.

In Section 27 the normal distribution was suggested as a tool for approximating probabilities from binomial experiments. The significance of the normal distribution goes much beyond this particular application. The normal distribution provides a means for computing relevant probabilities for many statistical inference procedures. In addition, experience shows that the normal distribution often furnishes a sufficiently accurate mathematical model to describe the distribution of many kinds of physical and industrial measurements, examination scores, and other types of quantitative data.

EXAMPLE 5.6 Instructors in large classes sometimes grade "on a curve." This means that course grades are assumed to follow a distribution that approximates a normal distribution as in Figure 5.7, where numerical grades

FIGURE 5.7

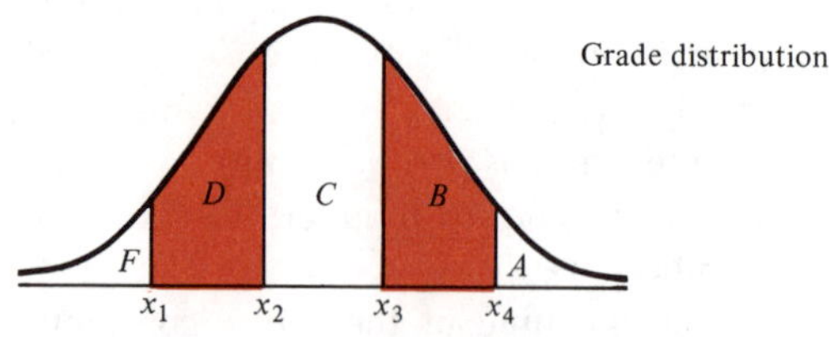

are measured along the horizontal axis. The dividing points x_1, x_2, x_3, and x_4 are frequently taken as $\mu - \frac{3}{2}\sigma$, $\mu - \frac{1}{2}\sigma$, $\mu + \frac{1}{2}\sigma$ and $\mu + \frac{3}{2}\sigma$, respectively, so that, for example, the proportion of *B*-grades in the course is made to equal the area under a normal curve between $\mu + \frac{1}{2}\sigma$ and $\mu + \frac{3}{2}\sigma$. We will see in Example 5.30 that this proportion approximately equals .24.

In view of the importance of the normal distribution, we now study how areas connected with normal curves are found.

Areas under the Normal Curve

29 Suppose that we have a quantity x, like grades in Example 5.6, normally distributed with mean μ and standard deviation σ, and are interested in the probability that x takes a value between two numbers x_1 and x_2. This probability $P(x_1 \leq x \leq x_2)$ is given by the area under the appropriate normal curve between x_1 and x_2 as indicated in Figure 5.8.

FIGURE 5.8

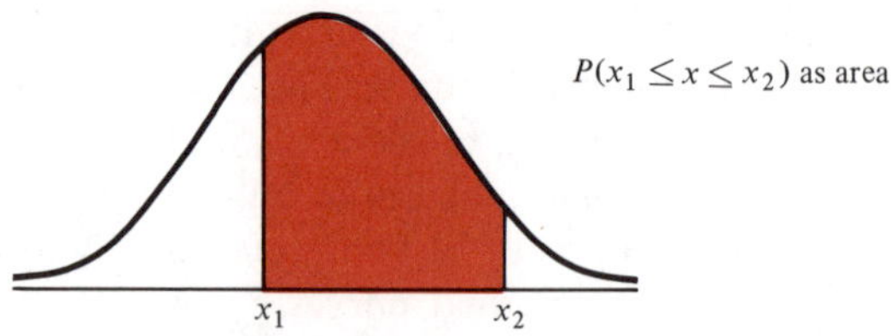

For the binomial distribution we need separate tables for each combination of n and p. A single table suffices for the determination of areas involving normal curves, whatever the appropriate mean and standard deviation. We first convert our original x-measurements into so-called standardized measurements z related to x as follows,

5.9
$$z = \frac{x - \mu}{\sigma}.$$

The points $\mu - \sigma$, μ and $\mu + \sigma$ on the x-scale of Figure 5.10 become

$$\frac{(\mu - \sigma) - \mu}{\sigma} = \frac{-\sigma}{\sigma} = -1,$$

$$\frac{\mu - \mu}{\sigma} = 0,$$

and $$\frac{(\mu + \sigma) - \mu}{\sigma} = \frac{\sigma}{\sigma} = +1,$$

indicating that on the z-scale we are dealing with the normal distribution with mean 0 and standard deviation 1. If we then set

$$z_1 = \frac{x_1 - \mu}{\sigma} \quad \text{and} \quad z_2 = \frac{x_2 - \mu}{\sigma},$$

Figure 5.10 shows that the area between x_1 and x_2 under the normal

FIGURE 5.10 Conversion to Standard Units

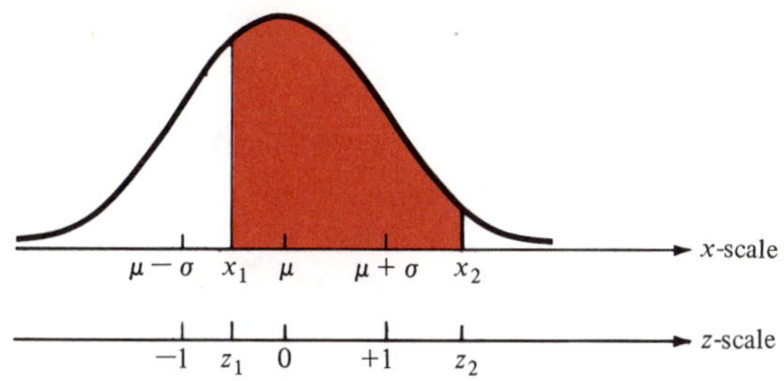

distribution with mean μ and standard deviation σ can be computed as the area between z_1 and z_2 under the normal curve with mean 0 and standard deviation 1. We will refer to the normal curve with mean 0 and standard deviation 1 as the *standard normal curve.*

30 Let z be a *positive* number. We denote the area under the standard normal curve between 0 and z by $A(z)$, and the area between $-z$ and 0 by $A(-z)$, as shown in Figures 5.11 and 5.12. Table B gives values of $A(z)$ (rounded off to four decimals) for values of z starting at $z = 0$ and increasing in steps of .01. Figures 5.11 and 5.12 show that, in view of the symmetry of the standard normal curve about $z = 0$, we have

$$A(-z) = A(z).$$

FIGURE 5.11

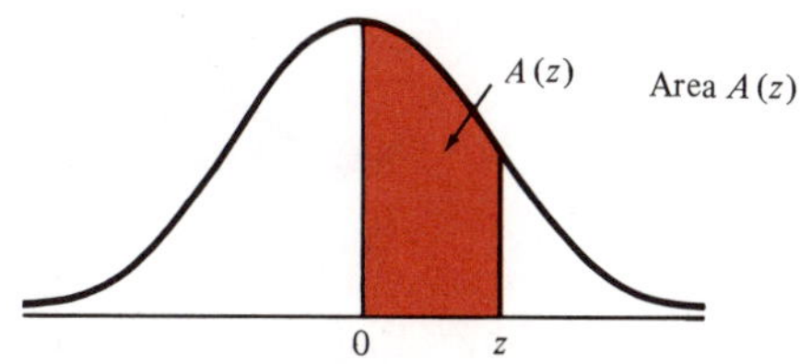

FIGURE 5.12

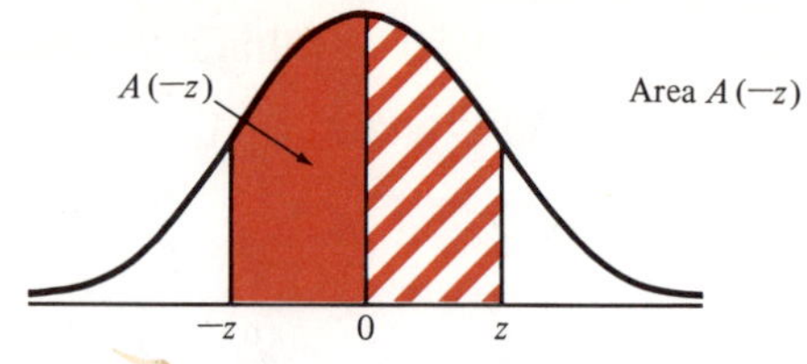

The problem of finding the area under the standard normal curve between two numbers z_1 and z_2 is now solved as is shown in the following examples. Since such areas represent probabilities, it is often useful and convenient to use the notation $P(z_1 \leq z \leq z_2)$, the probability that the standard normal variable z takes a value between z_1 and z_2, when referring to such areas.

EXAMPLE 5.13 Find the area under the standard normal curve between 0.50 and 1.50. Since the area between 0.50 and 1.50, that is, $P(0.50 \leq z \leq 1.50)$, equals the area between 0 and 1.50 minus the area between 0 and 0.50 (see Figure 5.14), we have

$$\begin{aligned} P(0.50 \leq z \leq 1.50) &= A(1.50) - A(0.50) \\ &= .4332 - .1915 = .2417. \end{aligned}$$

FIGURE 5.14

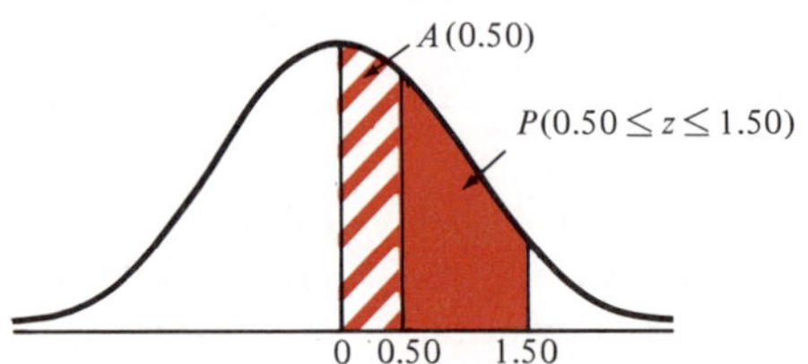

EXAMPLE 5.15 Find the area under the standard normal curve between -2 and -1. Figure 5.16 shows that, in view of the symmetry of the standard normal curve about 0, the area between -2 and -1 equals the area between 1 and 2, so that

$$\begin{aligned} P(-2 \leq z \leq -1) = P(1 \leq z \leq 2) &= A(2) - A(1) \\ &= .4772 - .3413 = .1359. \end{aligned}$$

FIGURE 5.16

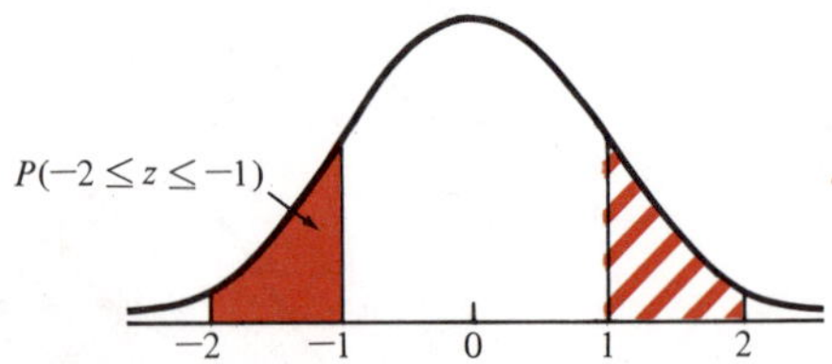

EXAMPLE 5.17 Find the area under the standard normal curve between -0.50 and 1. Figure 5.18 shows that the required area equals the sum of the area

between -0.50 and 0 and the area between 0 and 1. Thus

$$P(-0.50 \leq z \leq 1) = A(-0.50) + A(1) = A(0.50) + A(1)$$
$$= .1915 + .3413 = .5328.$$

FIGURE 5.18

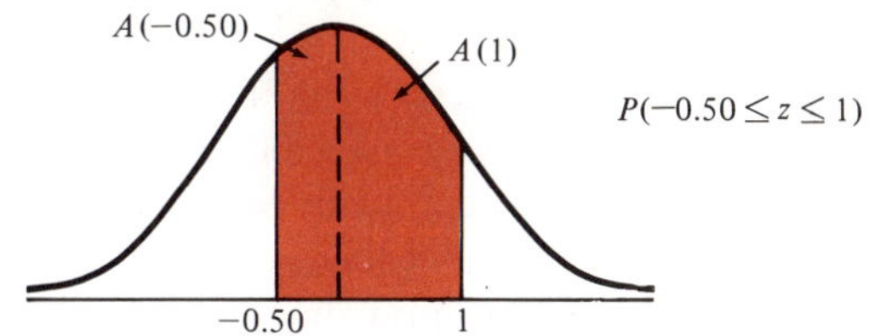

31 Statisticians frequently use so-called *tail* probabilities, probabilities like $P(z > 1)$ or $P(z < -2)$. We therefore define a function $T(z)$ such that for any *positive* number z, $T(z)$ equals the area under the standard normal curve to the right of z, while $T(-z)$ equals the area under the standard normal curve to the left of $-z$ as illustrated by Figure 5.19.

Since the standard normal curve is symmetric about 0 and, like all distributions, encloses total area 1, Figure 5.19 shows that $A(z) + T(z) = \frac{1}{2}$, so that

$$T(z) = \frac{1}{2} - A(z).$$

FIGURE 5.19

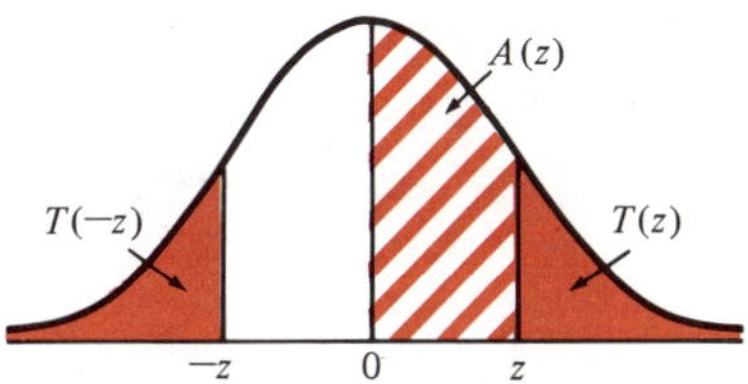

Table B lists $T(z)$ in addition to $A(z)$.

EXAMPLE 5.20

$$P(z < -2) = T(-2) = T(2) = .0228.$$

32 In statistical applications, we often have a reverse problem. Rather than find the value of $A(z)$ or $T(z)$ for given z, we need to find the number z that corresponds to a given probability $A(z)$ or $T(z)$.

EXAMPLE 5.21 Find the number that corresponds to a lower tail probability of .025. Figure 5.19 shows that the desired number is the negative of the number z such that $T(z) = .025$. From Table B we find that $T(1.96) = .025$. Thus the desired number is -1.96.

EXAMPLE 5.22 Find the number z such that $T(z) = T(-z) = .10$. According to Table B, $T(1.28) = .1003$ and $T(1.29) = .0985$. Thus the desired number z lies between 1.28 and 1.29. For our purposes, it is usually sufficiently accurate to determine z to two-place accuracy. We would then use $z = 1.28$, since .10 is closer to .1003 than to .0985.

EXAMPLE 5.23 Find the number z such that the area under the standard normal curve between $-z$ and z equals .60. Figure 5.24 shows that this problem is equivalent to the problem of determining the value z such that $A(z) + A(-z) = .60$. But then $A(z) = .60/2 = .30$ and, from Table B, we find that z equals approximately 0.84.

FIGURE 5.24

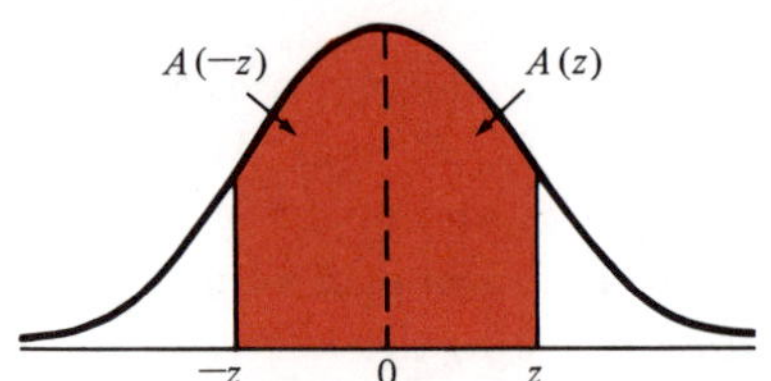

Problems similar to the ones just discussed arise so frequently in statistics that it is convenient to have a separate table giving values of z corresponding to frequently used probability levels. Table C provides this information. For reasons that will become evident in subsequent chapters, in Table C we write α' for either tail probability, $T(z)$ or $T(-z)$, α'' for the sum of these two tail probabilities, and γ for the central area between $-z$ and z as in Figure 5.25:

$$\alpha' = T(z) = T(-z);$$
$$\alpha'' = T(z) + T(-z);$$
$$\gamma = A(z) + A(-z).$$

FIGURE 5.25

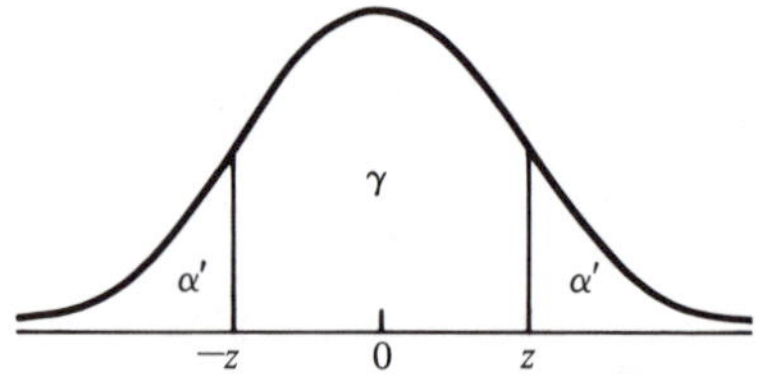

Then we can derive the following relationships between α', α'', and γ:

$$\alpha'' = 2\alpha' \qquad \alpha'' = 1 - \gamma \qquad \alpha' = \frac{1}{2}(1 - \gamma).$$

EXAMPLE 5.26 The problem in Example 5.22 is solved faster and more accurately by finding $\alpha' = .10$ in the third column of Table C and reading the z-value in the fourth column of the same row, $z = 1.282$. Similarly, Example

5.23 is solved by finding $\gamma = .60$ in the first column of Table C and reading the corresponding z-value, $z = 0.842$.

33 We will now illustrate some uses of Tables B and C in the solution of problems involving general normal distributions.

EXAMPLE 5.27 Scores on a widely used examination are known to follow a normal distribution with mean 500 and standard deviation 100.

a. What is the probability that a student who takes this examination receives a score between 300 and 600? The desired probability is given by the area—call it A—between $x_1 = 300$ and $x_2 = 600$ under the normal curve with mean 500 and standard deviation 100. According to our discussion in Section 29, A equals the area under the *standard* normal curve between

$$z_1 = \frac{x_1 - \mu}{\sigma} = \frac{300 - 500}{100} = -2$$

and

$$z_2 = \frac{x_2 - \mu}{\sigma} = \frac{600 - 500}{100} = 1.$$

We then have, similar to Example 5.17,

$$\begin{aligned} A = A(-2) + A(1) &= A(2) + A(1) \\ &= .4772 + .3413 = .8185. \end{aligned}$$

b. What is the probability of scoring 300 or less? This probability is given by the area to the left of 300 under the normal curve with mean 500 and standard deviation 100. Equivalently, it equals the area under the standard normal curve to the left of

$$z = \frac{300 - 500}{100} = -2.$$

This area is $T(-2) = .0228$.

c. Find the 90th percentile of all examination scores. The 90th percentile score has the property that 90 percent of all students who take the examination have lower scores, 10 percent have higher scores. Thus we want to find the value x such that the area to the right of x under the normal curve with mean 500 and standard deviation 100 equals .10. From Table C we find that $z = 1.282$ corresponds to an upper tail probability of .10. According to (5.9) of Section 29, x- and z-values are related as $z = (x - \mu)/\sigma$. But then

5.28 $$x = \mu + z\sigma,$$

and we find that $x = 500 + 1.282 \times 100 = 628.2$.

d. Find the numbers x_1 and x_2 that contain the central 75 percent of examination scores. Entering Table C with $\gamma = .75$, we find that

$z = 1.15$. Thus the central 75 percent of the area under the standard normal curve lies between -1.15 and 1.15, so that by (5.28), $x_1 = \mu - 1.15\sigma = 500 - 115 = 385$ and $x_2 = \mu + 1.15\sigma = 500 + 115 = 615$.

Figure 5.29 illustrates the following generalization of Example 5.27d.

FIGURE 5.29

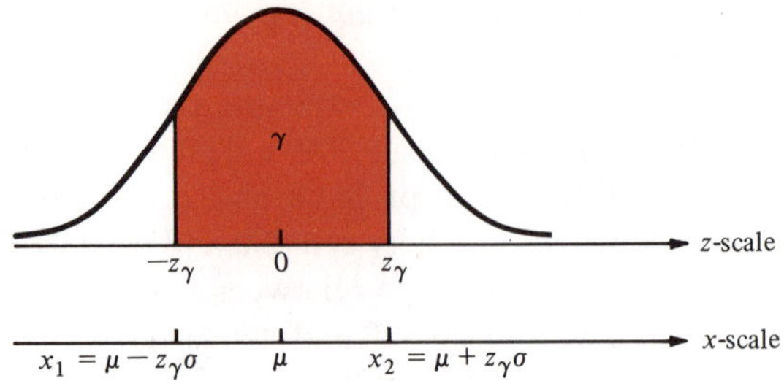

The central 100γ percent of the area under a normal distribution with mean μ and standard deviation σ lies between $\mu - z_\gamma\sigma$ and $\mu + z_\gamma\sigma$, where z_γ is the z-value that goes with γ in Table C. When we compute $z = (x-\mu)/\sigma$, we find out how many standard deviations the point x is away from the mean μ.

EXAMPLE 5.30 Find the proportion of students who receive grades of *A, B, C, D,* and *F* when grading on a curve according to the scheme in Example 5.6. These proportions are equal to the areas p_A, p_B, p_C, p_D, and p_F in Figure 5.31,

FIGURE 5.31

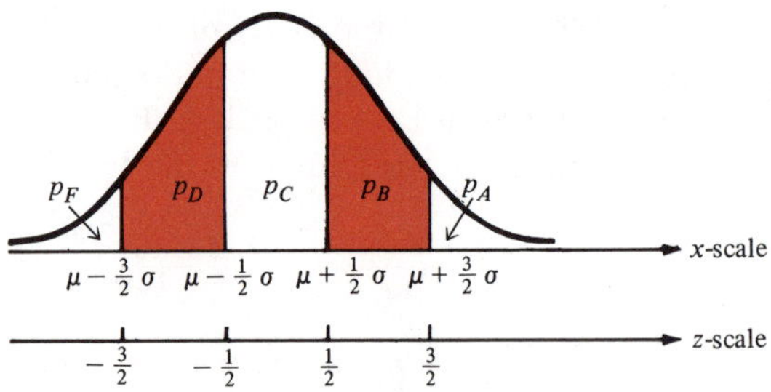

corresponding to dividing points $x_1 = \mu - \frac{3}{2}\sigma$, $x_2 = \mu - \frac{1}{2}\sigma$, $x_3 = \mu + \frac{1}{2}\sigma$, and $x_4 = \mu + \frac{3}{2}\sigma$ under the normal curve with mean μ and standard deviation σ. The corresponding standardized dividing points on the z-scale are

$$z_1 = \frac{x_1 - \mu}{\sigma} = \frac{(\mu - \frac{3}{2}\sigma) - \mu}{\sigma} = \frac{-\frac{3}{2}\sigma}{\sigma} = -\frac{3}{2}$$

and, by similar computations, $z_2 = -\frac{1}{2}$, $z_3 = \frac{1}{2}$, and $z_4 = \frac{3}{2}$. Thus $p_A = T(\frac{3}{2}) = .0668$ and $p_B = A(\frac{3}{2}) - A(\frac{1}{2}) = .4332 - .1915 = .2417$. By symmetry, $p_F = p_A = .0668$ and $p_D = p_B = .2417$. Finally, $p_C = 1 - (p_A + p_B + p_D + p_F) = .3830$. Note that these proportions do not depend on the actual values of μ and σ of the normal curve that describes the distribution of grades. When this system of grading on a curve is used, the top 7 percent of students receive grades of A, the next 24 percent receive grades of B, the central 38 percent receive grades of C, followed by 24 percent of D's and 7 percent of F's.

Normal Approximation for Binomial Probabilities

34 Now we finally return to the problem that led us in the first place to discuss the normal distribution. How can we approximate the probability of observing between 10 and 25 successes in 48 binomial trials with success probability $1/3$? In Section 27 we proposed to approximate such probabilities by means of areas under an appropriate normal curve. It can be shown (see Example A.13 in the Appendix) that for binomial probabilities involving n trials with success probability p, the appropriate mean and standard deviation are $\mu = np$ and $\sigma = \sqrt{npq}$. In particular, for $n = 48$ and $p = 1/3$, we find that $\mu = 48 \times 1/3 = 16$ and $\sigma = \sqrt{48 \times 1/3 \times 2/3} = 3.266$. Since we want the probability of observing at least 10 and no more than 25 successes in our 48 trials, we find the area between 10 and 25 under the normal curve with mean 16 and standard deviation 3.266. The corresponding standardized values on the z-scale are

$$z_1 = \frac{10 - 16}{3.266} = -1.84 \qquad \text{and} \qquad z_2 = \frac{25 - 16}{3.266} = 2.76,$$

so that similar to Example 5.17 the required area equals $A(-1.84) + A(2.76) = .964$.

35 This procedure can be generalized. Let k_1 and k_2 be two integers satisfying the condition $0 \leq k_1 < k_2 \leq n$. We should like to approximate the probability that, in n binomial trials with success probability p, we observe between k_1 and k_2 successes, both limits included. Our previous discussion suggests that if the number of trials is sufficiently large, the area under the normal distribution with mean $\mu = np$ and standard deviation $\sigma = \sqrt{npq}$ between the points k_1 and k_2 provides a suitable approximation. Thus we have to compute the two standardized values

$$z_1 = \frac{k_1 - np}{\sqrt{npq}} \qquad \text{and} \qquad z_2 = \frac{k_2 - np}{\sqrt{npq}}$$

and find the area between z_1 and z_2 from Table B.

For most practical purposes the normal approximation will give sufficiently accurate results provided that npq is at least 3, that is, provided that $n \geq 3/pq$. In particular, if the success probability p is near $\frac{1}{2}$, the normal approximation may be used for n as small as 12. However, if p is near 0 or 1, the number of trials should be considerably larger before we rely on the normal approximation. Thus for $p = \frac{1}{10}$, the preceding rule suggests that we have at least $3/(.1)(.9) = 300/9 = 33$ trials.

EXAMPLE 5.32 Use the normal approximation to compute the probability of observing at least 80 and no more than 90 successes in 400 trials with success probability $\frac{1}{4}$. Since

$$np = 400 \times \tfrac{1}{4} = 100$$

and $$\sqrt{npq} = \sqrt{400 \times \tfrac{1}{4} \times \tfrac{3}{4}} = \sqrt{75} = 8.66,$$

we find that $z_1 = (80 - 100)/8.66 = -2.31$ and $z_2 = (90 - 100)/8.66 = -1.15$, and, as in Example 5.15, the desired approximation equals $A(2.31) - A(1.15) = .115$.

If $k_1 = 0$, it is more accurate, and less work, to replace z_1 by minus infinity, so that the probability of observing no more than k_2 successes is approximated by the area under the standard normal curve to the left of $z_2 = (k_2 - np)/\sqrt{npq}$. Similarly, if $k_2 = n$, z_2 is best replaced by plus infinity, so that the probability of observing at least k_1 successes is approximated by the area under the standard normal curve to the right of $z_1 = (k_1 - np)/\sqrt{npq}$.

EXAMPLE 5.33 Use the normal approximation to compute the probability that, in 150 trials with success probability .6, we observe at least 100 successes. In this problem we have $k_1 = 100$ and $z_1 = (100 - 150 \times .6)/\sqrt{150 \times .6 \times .4} = (100 - 90)/\sqrt{36} = 1.67$. Since $k_2 = n$, we take z_2 as plus infinity, and the desired approximation equals $T(1.67) = .048$.

36* *A Technical Note*

The accuracy of the normal approximation for binomial probabilities can usually be improved by a device called a *continuity correction.* The nature of this device is best illustrated by returning once more to the problem of approximating the probability of observing at least 10 and no more than 25 successes discussed earlier. The exact probability is given by $b(10) + \cdot \cdot \cdot + b(25)$ and is represented graphically by a corresponding sum of rectangles as in Figure 5.34. When we approximate this sum of rectangles by the area under the appropriate normal curve between $k_1 = 10$ and $k_2 = 25$, we ignore the shaded areas in Figure 5.34. This suggests that possibly a better approximation is obtained by using the area between $x_1 = k_1 - \frac{1}{2} = 9.5$ and $x_2 = k_2 + \frac{1}{2}$

FIGURE 5.34

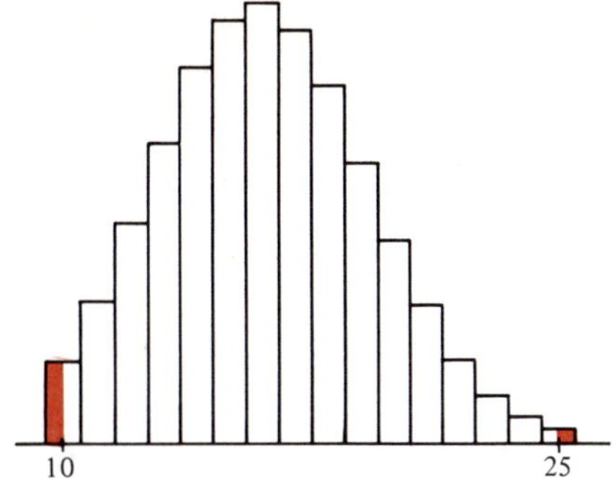

$= 25.5$. We then find $z_1 = -1.99$ and $z_2 = 2.91$, resulting in the approximation $A(-1.99) + A(2.91) = .975$. This value is closer to the true value .978 than the earlier approximation .964.

The quantity $\frac{1}{2}$ which is subtracted from k_1 and added to k_2 is known as a *continuity correction.*

37 Another example will illustrate the normal approximation. Suppose that we toss a fair coin 100 times, or, what amounts to the same, drop 100 fair coins on a table. In an experiment of this type, we might expect to get 50 heads and 50 tails. But exact computations reveal that the probability of an even split is only .08; 92 percent of the time we get either more than or fewer than 50 heads. We may then ask between what limits the number of heads is likely to fluctuate. In this form the question is too vague for an answer. Strictly speaking, the occurrence of 0 heads or 100 heads is not impossible, only extremely unlikely. We should ask, instead, how large a deviation from an even split can reasonably be expected. Of course, the word "reasonable" is also open to all sorts of interpretations, but statisticians are often willing to consider as reasonable deviations that occur about 95 percent of the time. Before you read on, you may want to guess what kind of deviations are reasonable in 100 tosses of a coin.

From Table C we find that the area between -2 and 2 under the normal curve is approximately .95 (.954 to be exact). The remark following Example 5.27d shows that the same is true for the points $x_1 = \mu - 2\sigma$ and $x_2 = \mu + 2\sigma$ for normal distributions with mean μ and standard deviation σ. If we apply this result to the binomial distribution, we find that with probability approximately .95 the observed number of successes falls between $np - 2\sqrt{npq}$ and $np + 2\sqrt{npq}$. In our case, $np = 50$ and $\sqrt{npq} = \sqrt{25} = 5$, so that it is reasonable to expect that the number of heads is somewhere between 40 and 60.

PROBLEMS

1 Find the area under the standard normal curve to the left of

a. 1.55;
b. -1.55;
c. $-.86$;
d. 4.25.

2 Find the area under the standard normal curve to the right of

a. 1.11;
b. $-.45$;
c. 3.75;
d. -3.75.

3 Find the area under the standard normal curve between

a. 1.28 and 1.96;
b. -1 and 2;
c. -1.15 and 1.15.

4 Find the area under the standard normal curve outside the interval

a. $-.85$ to .85;
b. -2 to 2.

5 Find the value of z such that the area under the standard normal curve to the left of z is

a. .719;
b. .66;
c. .008;
d. .34.

6 Find the value of z such that the area under the standard normal curve to the right of z is

a. .30;
b. .78.

7 Find the value of z such that the area under the standard normal curve between $-z$ and z is

a. .90;
b. .68;
c. .34.

8 For the normal distribution with mean 500 and standard deviation 100, find

a. the area to the left of 450;
b. the area to the right of 650;
c. the area between 550 and 650;

d. the numbers that contain the central 90 percent of the area,
e. the numbers that contain the central 95 percent;
f. the value x such that the area to the right of x equals .05;
g. the value x such that the area to the left of x equals .05.

9 Heights of men in a certain population are normally distributed with mean $68\frac{1}{2}$ inches and standard deviation $2\frac{1}{2}$ inches. Find the proportion of men in this population with heights

a. less than 60 inches;
b. more than 72 inches;
c. between 60 and 72 inches.

10 Assume that the amount of time required to complete an examination is normally distributed with mean 100 minutes and standard deviation 15 minutes.

a. If 200 students take the examination, how many can be expected to complete the examination within two hours? (*Hint:* What proportion of students can be expected to finish within two hours?)
b. How much time should be allowed for the examination if it is desired that only 75 percent of the students who take the examination complete all parts?

11 Packages of cereal state that the net weight is 16 oz. From past experience, the company knows that actual net weights are normally distributed with standard deviation $\frac{1}{2}$ oz. How should the filling mechanism be adjusted so that no more than 1 package in 100 has a net weight that is less than 16 oz? (*Hint:* Find the mean μ of the normal distribution with standard deviation .5 that satisfies Figure 5.35.)

FIGURE 5.35

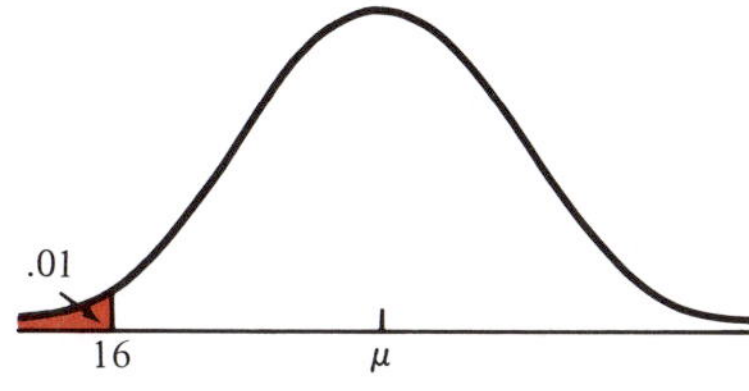

12 We observe the results of 225 binomial trials with success probability .20. Use the normal approximation to compute the following probabilities:

a. $P(40 \leq k \leq 50)$;
b. $P(k \geq 55)$;
c. $P(k \leq 60)$;
d. $P(42 \leq k \leq 58)$.

13 Use the normal approximation to find the following binomial probabilities:

a. $P(20 \leq k \leq 30)$, if $n = 80$ and $p = .30$;
b. $P(k \leq 40)$, if $n = 100$ and $p = .40$;

c. $P(50 \leq k \leq 60)$, if $n = 80$ and $p = .70$;
d. $P(k \geq 55)$, if $n = 300$ and $p = .25$.

14 If a fair coin is tossed 200 times, what are "reasonable" limits for the number of heads? (*Hint:* See the discussion in Section 37.)

15 Find the following probabilities relating to binomial distributions with success probability $p = \frac{1}{2}$:

a. $P(40 \leq k \leq 60)$ if $n = 100$;
b. $P(90 \leq k \leq 110)$ if $n = 200$;
c. $P(190 \leq k \leq 210)$ if $n = 400$;
d. $P(180 \leq k \leq 200)$ if $n = 400$.

16 Past experience indicates that 60 percent of the passengers on an early morning flight request hot breakfast, while the remaining 40 percent prefer cold breakfast. For each flight, the cabin crew has on hand 72 hot breakfasts and 48 cold breakfasts. If 100 passengers take the flight, what is the probability that every passenger receives the desired breakfast? (*Hint:* The cabin crew will not be able to satisfy the requests of all passengers, if more than 72 or fewer than 52 passengers request hot breakfast.)

6

ESTIMATION

38 In Chapter 4 we discussed a problem of the following type: An experiment consists of performing n independent trials, each with two possible outcomes, success and failure, with constant known success probability p. What is the probability that we observe some given number of successes? For example, what is the probability that we observe 12 successes in 48 trials with success probability $\frac{1}{3}$? This is a problem in probability.

Now consider a different setup. We have actually observed the results of 48 trials of the type just described, say, 12 successes and 36 failures. But we do not know the value of the success probability p that brought about these results. We want to use the available information to estimate the unknown value of p. This is a statistical problem.

A quantity like p that is characteristic of the given experiment is called a parameter. Other examples of parameters are the mean μ and the standard deviation σ of a normal distribution. In this chapter we will discuss methods of estimating the parameter p of a binomial distribution.

A Point Estimate for p

39 In the preceding example most readers would presumably decide to use the observed relative frequency of success, $\frac{12}{48} = \frac{1}{4}$, to estimate p. Other estimates are possible. However, in this book we will consider only this particular estimate. Thus if in n independent trials with constant (unknown) success probability p we observe k successes and $n - k$ failures, we will estimate p as the relative frequency of success, k/n.

How good is this estimate? It is implied in the frequency interpretation of probability that if the number of trials is sufficiently large, the relative frequency of success (and therefore our estimate of p) can be expected to differ arbitrarily little from the success probability p. The question is, how large is sufficiently large? There is no unique answer to this question. Whenever we generalize from sample information, we must be prepared for erroneous conclusions unless we are satisfied with trivial statements. The best we can hope for are statements that have a certain known probability of being correct. Thus we should determine the probability that our estimate does not deviate from the true success probability p by more than some positive quantity, say ε. The following illustrates the situation:

$\frac{k}{n}$

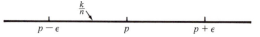

$p-\epsilon$ $\quad p \quad$ $p+\epsilon$

In mathematical language, we want to know the probability, call it $P(\varepsilon)$, that

$$p - \varepsilon \leq \frac{k}{n} \leq p + \varepsilon.$$

We can convert this statement into a more familiar one by multiplying all three parts by n. We then see that $P(\varepsilon)$ equals the probability of the event

$$np - n\varepsilon \leq k \leq np + n\varepsilon,$$

that is, the probability that, in n trials with success probability p, the number of successes lies between $np - n\varepsilon$ and $np + n\varepsilon$. If n is sufficiently large, this probability can be evaluated using the normal approximation of Section 35. The appropriate standardized z-values for the approximation are

$$z_1 = \frac{np - n\varepsilon - np}{\sqrt{npq}} = \frac{-n\varepsilon}{\sqrt{npq}} = \frac{-\varepsilon\sqrt{n}}{\sqrt{pq}}$$

and
$$z_2 = \frac{np + n\varepsilon - np}{\sqrt{npq}} = \frac{n\varepsilon}{\sqrt{npq}} = \frac{\varepsilon\sqrt{n}}{\sqrt{pq}},$$

so that $P(\varepsilon)$ is given by the area under the standard normal curve between $-z$ and z with $z = \varepsilon\sqrt{n}/\sqrt{pq}$ as indicated in Figure 6.1. It follows that

$$P(\varepsilon) \doteq 2A\left(\frac{\varepsilon\sqrt{n}}{\sqrt{pq}}\right),$$

where the symbol $\doteq$ indicates that the left and right sides are not exactly, but only approximately equal.

This result poses something of a dilemma. To compute $P(\varepsilon)$, we have to know the value of $z = \varepsilon\sqrt{n}/\sqrt{pq}$ which depends on the unknown success probability p that we want to estimate. We seem to be going

FIGURE 6.1

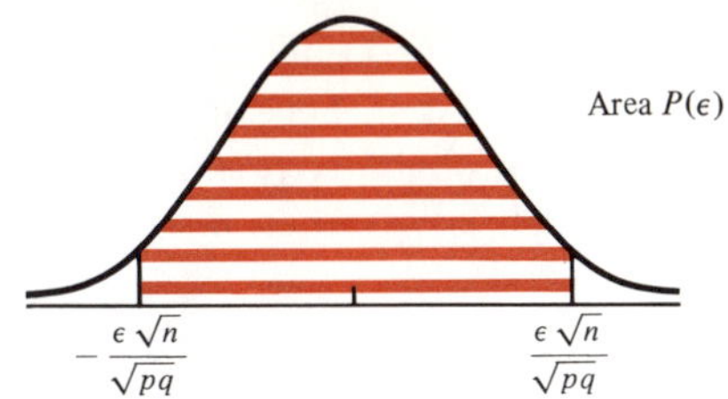

in circles. Yet let us not give up. The quantity that enters into the determination of z is $\sqrt{pq}$. Table 6.2 evaluates $\sqrt{pq}$ (as well as pq) for various values of p. We see that $\sqrt{pq}$ is never larger than $1/2$ and progressively decreases in value as p deviates more and more from .5. Since $z = \varepsilon\sqrt{n}/\sqrt{pq}$, it follows that z cannot be smaller than $2\varepsilon\sqrt{n}$. Whatever the true success probability p, we can be sure that

$$P(\varepsilon) \geq 2A(2\varepsilon\sqrt{n}),$$

which is represented by the solidly shaded area in Figure 6.3.

TABLE 6.2

p	.5	.4 or .6	.3 or .7	.2 or .8	.1 or .9	.05 or .95
pq	.25	.24	.21	.16	.09	.0475
$\sqrt{pq}$	.50	.49	.46	.40	.30	.22

FIGURE 6.3

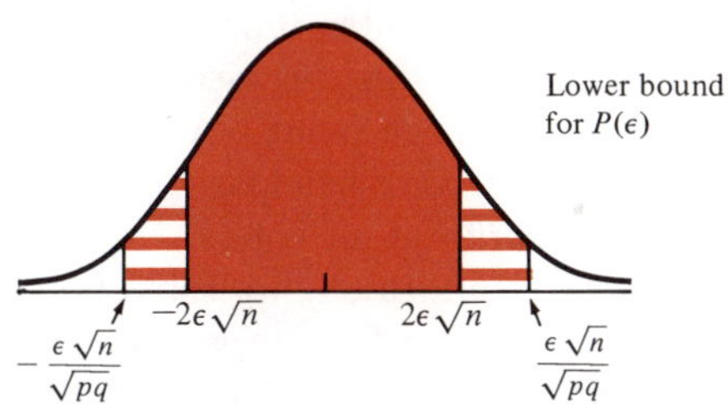

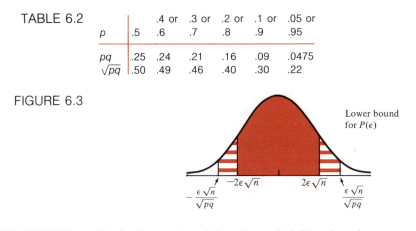

EXAMPLE 6.4 Find a lower bound for the probability that the estimate k/n of the success probability p based on the results of 100 trials does not deviate from the true value p by more than .05. We have $n = 100$, $\varepsilon = .05 = 1/20$, so that $2\varepsilon\sqrt{n} = 1$ and $P(\varepsilon) \geq 2A(1) = .68$.

40 We can now answer the question of how many observations are needed to give a satisfactory estimate for the success probability p. Given ε, we want to be reasonably sure that the resulting estimate does not deviate from the true value p by more than ε. Here "reasonably sure" is interpreted to mean with a given high probability, say, with a probability of at least .95. Using our previous terminology, we want $P(\varepsilon) \geq .95$. Since the area under the standard normal curve between -2 and 2 equals .954 according to Table C, Figure 6.1 shows that our requirement is certainly satisfied if $\varepsilon\sqrt{n}/\sqrt{pq} = 2$ or, solving for n, $n = 4pq/\varepsilon^2$. As in Section 39 we can get around the difficulty that the success probability p is unknown by replacing pq by its maximum value, which is $\frac{1}{4}$. Then $n = 1/\varepsilon^2$, which depends only on ε.

TABLE 6.5

ε	.10	.05	.02	.01
$n = 1/\varepsilon^2$	100	400	2,500	10,000

Table 6.5 evaluates $1/\varepsilon^2$ for some typical values of ε. We see that if we do not mind an error of .05, 400 trials are sufficient. But we need 10,000 trials if we want to be reasonably certain that our estimate is in error by at most .01.

Table 6.5 can also be used in reverse. Suppose that we have decided to observe the results of 100 trials and then estimate the unknown success probability p as the relative frequency k/n. Table 6.5 tells us that we can be reasonably sure that this estimate is not going to be in error by more than .10.

An interesting application of the preceding results is to public opinion polls such as those conducted before presidential elections. While the sampling scheme underlying such polls is considerably more complicated than the one we have discussed, our formula still can be applied to give some indication of the accuracy of a poll. Suppose that our poll is based on 2500 interviews. (Actually most polls use fewer.) Table 6.5 tells us that an estimate based on such a poll may be in error by as much as two percentage points. A candidate who gets only 48.5 percent of the votes of those participating in a poll may nevertheless be the choice of the majority of all voters.

Let us return briefly to Table 6.2. We can see that for p between .3 and .7 the product pq is not much smaller than .25. If p can be expected to be in this interval (and the statistician usually has such rough information), we essentially need $n = 1/\varepsilon^2$ trials to produce a satisfactory estimate. On the other hand, if p is close to 0 or 1, substantially fewer trials are sufficient. Thus, for instance, if we have reason to think that the true value of p is smaller than $\frac{1}{10}$ (or greater than $\frac{9}{10}$), it is sufficient to use only

$$n = 4 \times \frac{1}{10} \times \frac{9}{10} \times \frac{1}{\varepsilon^2} = \frac{.36}{\varepsilon^2}$$

trials, a saving of 64 percent.

41 When planning a statistical investigation, we often confront the question of how many more observations we would need to double our accuracy. Of course, we have to explain what we mean by doubling our accuracy. In the present case a very simple interpretation is possible. Since we must be prepared to commit a certain error in estimating p, doubling our accuracy should be interpreted as cutting this possible error in half. Since ε is a measure of the possible error, this means cutting ε in half. It may seem that we should be able to do this by taking twice as many observations, but let us look again at our formula determining n as a function of ε, $n = 4pq/\varepsilon^2$. If we replace ε by $\varepsilon/2$, we find that

$$n = \frac{4pq}{\varepsilon^2/4} = \frac{16pq}{\varepsilon^2}.$$

The new n must be four times as large as the old n. Since, in general, it takes time and money to provide additional observations, there comes a point when costs of additional observations more than outweigh any possible gain in greater accuracy. This is a typical example of the law of diminishing returns. If we have 100 observations, 300 additional observations double our accuracy. But if we already have 1000 observations, 300 additional observations do very little to increase our accuracy.

The relationship that we have just discovered between the number of observations and the resulting accuracy is not limited to the problem of estimating the parameter p of a binomial distribution. It arises over and over again in other statistical problems. Increasing the accuracy of a statistical procedure by a factor of 2 usually requires four times as many observations; tripling the accuracy requires nine times as many observations. And a ten-fold increase in accuracy can be achieved only by a 100-fold increase in the number of observations. Indeed, such an increase in the number of observations may well be accompanied by complications that would upset any expected gain in accuracy.

42 The estimate k/n of the success probability p is called a *point estimate.* The reason for this name is rather obvious. The value k/n represents a single point in the interval from 0 to 1 of all possible p-values. In practical applications point estimates are used more frequently than any other kind of statistical procedure. But point estimates do have their limitations.

Consider an experiment where we observe 55 successes in 100 trials. Now suppose that we observe 5500 successes in another experiment involving 10000 trials. In both cases the point estimate of p is .55 and there is no way of distinguishing the second estimate from the first. Nevertheless, our earlier considerations imply that the accuracy of the second estimate is ten times as great as the accuracy of the first estimate. Unless we specifically add a statement that indicates in some way the degree of accuracy of a given point estimate, we do not know whether the estimate is highly reliable or is not even worthy of being called an estimate. In the past when a public opinion poll stated that 55 percent of the people preferred proposition A to proposition B, did you ever wonder by how much this statement might be in error? Public opinion polls rarely, if ever, mention how accurate their estimates are.

Interval Estimates

43 Let us look again at Table 6.5 relating n and ε. We see from this table that a point estimate based on 100 observations may reasonably be expected to be in error by no more than .10, while the corresponding error of an estimate based on 10000 observations is .01. Assume that both times the point estimate is .55. In the first case the true value of p would then be somewhere between $.55 - .10 = .45$ and $.55 + .10 = .65$. In the second case the corresponding two values are $.55 - .01 = .54$ and $.55 + .01 = .56$. We will call such an interval of possible p-values a confidence interval for the parameter p. In our specific example we have the following picture:

n	*k*	*Point Estimate*	*Confidence Interval*
100	55	.55	$.45 \leq p \leq .65$
10000	5500	.55	$.54 \leq p \leq .56$

The much shorter length of the second interval clearly shows the superiority of the second estimate. Indeed, the length of the second interval is $.56 - .54 = .02$, while that of the first interval is $.65 - .45 = .20$. The length of the first interval is ten times that of the second interval showing immediately the ten-fold increase in accuracy as we go from 100 to 10000 trials. Unless a problem of statistical estimation specifically requires a point estimate as an answer, interval estimates are usually preferable. We will now investigate interval estimates in greater detail.

44 How can we formulate the above procedure in more general terms? In Section 39 we investigated the probability $P(\varepsilon)$ that the point estimate k/n did not deviate from the true success probability p by more than ε. Now whenever k/n is within distance ε of p, it is also true that p is within distance ε of k/n, or

6.6 $$\frac{k}{n} - \varepsilon \leq p \leq \frac{k}{n} + \varepsilon,$$

as is illustrated in the following picture:

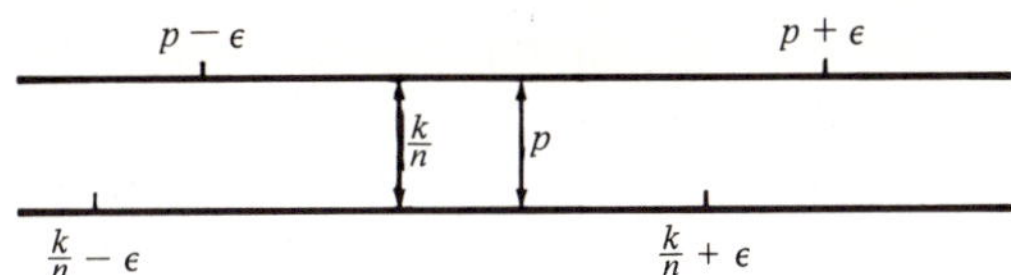

We will call (6.6) a confidence interval for the parameter p with *confidence coefficient* $\gamma = P(\varepsilon)$. In our earlier investigations we assumed that ε was given, but with confidence intervals it is more appropriate for the statistician to fix the confidence coefficient γ and to determine the value of ε correspondingly. Before we investigate this problem, we will discuss the meaning of the confidence coefficient γ and its selection.

45 Theoretically, the unknown success probability p may have any value between 0 and 1. When we use interval (6.6), we imply that we are "confident" that on the basis of available information, namely, k successes in n trials, the true value of the success probability p really lies in the shorter interval bounded by

$$\frac{k}{n} - \varepsilon \qquad \text{and} \qquad \frac{k}{n} + \varepsilon.$$

Statisticians can never be 100 percent certain of their statements, only reasonably certain. The confidence coefficient associated with a confidence interval is a measure of the degree of confidence that a statistician has in the correctness of a statement. To be more precise, suppose that the statistician has decided on the confidence coefficient .95. Confidence intervals with confidence coefficient .95 contain the true but unknown value of the success probability 95 percent of the time. On the other hand, 5 percent of the time such intervals computed on the basis of available sample information do not contain the true value p. Of course, in any particular case the statistician does not know whether the first or the second situation prevails. All the statistician knows is that before starting an investigation there is probability .95, or, more generally, probability γ, that the observed number of successes k resulting from the experiment will give an interval (6.6) that actually contains the true success probability p. (See Problem 17.)

Why should we be satisfied with a confidence interval that has confidence coefficient .90 if we can have one with confidence coefficient .95? Or why choose $\gamma = .95$, if we can choose $\gamma = .99$? The answer is that we want not only a reliable, but also an informative, statement. An interval stating that p lies between .2 and .9 has little practical value. This kind of rough information is usually known to the investigator without experimentation.

As we will see in Section 46, increasing the confidence coefficient γ also increases the length of the resulting confidence interval (assuming that other circumstances remain unchanged). Correspondingly, a reduction in the length of a confidence interval can be achieved only at the expense of a reduced degree of confidence in the correctness of the interval. We have to compromise our demands for shortness of the interval on the one hand and reliability on the other hand. Now we return to our discussion of the determination of ε.

46 We have seen in Figure 6.1 that $P(\varepsilon) \doteq 2A(z)$, where $z = \varepsilon\sqrt{n}/\sqrt{pq}$. If we then select z in such a way that $2A(z) = \gamma$ and express ε in terms of z, the resulting confidence interval will have confidence coefficient approximately γ. We find that $\varepsilon = z\sqrt{pq}/\sqrt{n}$, giving the confidence interval

6.7
$$\frac{k}{n} - z\sqrt{\frac{pq}{n}} \leq p \leq \frac{k}{n} + z\sqrt{\frac{pq}{n}}.$$

Since $2A(z)$ equals the area under the standard normal curve between $-z$ and z, the z-value, z_γ, that produces a confidence interval with confidence coefficient γ can be read directly from Table C.

We now see the effect of changing the confidence coefficient on the confidence interval. According to (6.7), the length of the confidence interval is $2z\sqrt{pq/n}$. Thus the length of the interval increases as z increases. Since z increases together with the confidence coefficient γ, a larger confidence coefficient results in a longer interval. Therefore, when deciding on a confidence coefficient, we should balance the advantages of a more reliable statement against the disadvantages of a longer interval.

In its present form the interval (6.7) cannot be computed, since the value of $\sqrt{pq}$ is unknown. One way around this difficulty is to replace $\sqrt{pq}$ by its maximum value $\frac{1}{2}$. This results in the interval

6.8
$$\frac{k}{n} - \frac{z}{2}\frac{1}{\sqrt{n}} \leq p \leq \frac{k}{n} + \frac{z}{2}\frac{1}{\sqrt{n}}.$$

In particular, for $z = 2$ corresponding to the confidence coefficient .95 (more accurately, .954), we have the interval

$$\frac{k}{n} - \frac{1}{\sqrt{n}} \leq p \leq \frac{k}{n} + \frac{1}{\sqrt{n}},$$

of which the two intervals discussed in Section 43 are special cases.

Unless p happens to have the value $\frac{1}{2}$, the interval (6.8) is actually longer than is necessary to achieve the *nominal* confidence coefficient corresponding to the value z used in computing the limits (6.8). Looking at it a different way, we can say that the true confidence coefficient associated with the interval (6.8) is really higher than we claim. Such an interval is said to be conservative. As long as the true value of p is close to $\frac{1}{2}$, there is little difference between the true and nominal confidence coefficients. However, if p is close to 0 or 1, the difference may be substantial and the conservative interval may not represent a very desirable solution. As an alternative, statisticians often use the estimate k/n for p in computing the value of $\sqrt{pq}$. With this substitution, the confidence interval (6.7) becomes

6.9 $$\frac{k}{n} - z\sqrt{\frac{(k/n)[1-(k/n)]}{n}} \leq p \leq \frac{k}{n} + z\sqrt{\frac{(k/n)[1-(k/n)]}{n}}.$$

EXAMPLE 6.10 Find a confidence interval with confidence coefficient .90 for the success probability p, when in 400 trials we observe 40 successes. Corresponding to $\gamma = .90$ in Table C, we find the value $z = 1.645$. Since

$$\frac{k}{n} = \frac{40}{400} = \frac{1}{10} = .10,$$

we have

$$z\sqrt{\frac{(k/n)[1-(k/n)]}{n}} = 1.645\sqrt{\frac{(.10)(.90)}{400}} = .025,$$

so that the confidence interval (6.9) becomes

$$.10 - .025 \leq p \leq .10 + .025$$

or

$$.075 \leq p \leq .125.$$

To determine the conservative interval (6.8), we compute

$$\frac{z}{2}\sqrt{\frac{1}{n}} = \frac{1.645}{2}\sqrt{\frac{1}{400}} = .041.$$

Thus we have the interval

$$.10 - .041 \leq p \leq .10 + .041$$

or

$$.059 \leq p \leq .141.$$

Comparing this with our previous result of $.075 \leq p \leq .125$, we see that here the conservative confidence interval is unnecessarily wide, .082 versus .050.

47 The need to determine a confidence interval for the parameter p of a binomial distribution arises so frequently that statisticians have devoted considerable attention to the problem. The methods discussed have the advantage of simplicity. However, tables and graphs exist that expedite finding confidence limits. For small values of n when the normal approximation to the binomial distribution is unsatisfactory, such tables are more accurate. For $n = 10$ and confidence coefficient at least .95, such a table might look like this:

TABLE 6.11 Confidence Intervals for p; $n = 10$, $\gamma \geq .95$

k	*Interval*
0	$0 \leq p \leq .31$
1	$0 \leq p \leq .45$
2	$.03 \leq p \leq .56$
3	$.07 \leq p \leq .65$
4	$.12 \leq p \leq .74$
5	$.19 \leq p \leq .81$
6	$.26 \leq p \leq .88$
7	$.35 \leq p \leq .93$
8	$.44 \leq p \leq .97$
9	$.55 \leq p \leq 1$
10	$.69 \leq p \leq 1$

Table A-23 in the book by Natrella cited in the Bibliography provides confidence intervals for samples up to size 30. For larger samples, Table A-24 of the same book provides useful graphs. Similar information is provided in Table 9.6 of Owen's *Handbook of Statistical Tables*.

48* *Randomized Response Method*

The methods of estimation discussed so far can be used to give answers to such questions as what proportion of students at a given college are heavy smokers. We select a random sample of students at the college and ask each student in the sample whether he smokes, say, at least 20 cigarettes a day. The relative frequency of yes-answers in the sample then constitutes an estimate of the proportion of heavy smokers in the student population. A confidence interval can be found by the methods of Section 46.

Presumably, no student is unwilling to answer a question concerned with cigarette smoking truthfully. The reaction may be quite different if the question were to involve drug use or sexual behavior. Confronted with a rather personal question, an individual may decide either not to answer at all or not to answer truthfully. Either choice introduces an undesirable bias into the survey. To cover such situations, statisticians have developed randomized response techniques to safeguard the respondent's privacy and still furnish the desired information.

As an example, consider a survey to determine the proportion of marijuana users. More exactly, it is desired to have a yes- or no-answer to the following question, which will be called question A from now on:

A: Have you smoked pot during the past week?

In a randomized response approach, the sensitive question (question A above) is paired with another completely harmless question, to be called question B. Examples are:

B_1: Was your father born in an even-numbered year?

Or if all participants have been instructed to toss a coin and watch whether it falls heads or tails, question B may be,

B_2: Did the coin fall tails?

The important thing is that the probability of a yes-answer to question B should be known. (For either question, B_1 or B_2, this probability is $\frac{1}{2}$.)

The actual survey is performed as follows. Each participant is instructed to answer either question A or question B depending on the outcome of some chance experiment with known probabilities. For example, each participant may be instructed to select a random digit from a table of random digits and answer question A if the digit should turn out to be 0, 1, 2, 3, 4, or 5 and answer question B if the digit should turn out to be 6, 7, 8, or 9. A participant will then answer question A with probability .6 and question B with probability .4. Of course, the interviewer should not know whether a participant is answering question A or question B. Under these conditions, a yes-answer does not imply that the interviewee is a marijuana smoker, nor does a no-answer imply that he or she is not. The interviewee's privacy is preserved.

The statistical problem we want to solve is how can we estimate the proportion of marijuana smokers from the sample answers? To answer this question, we have to solve first a probabilistic problem. What is the probability, say Π, that a randomly selected person will answer yes to the question as assigned by the random device? The answer is given by (3.20),

$$\Pi = P(E) = P(A)P_A(E) + [1 - P(A)]P_B(E).$$

Here $P(A)$ is the probability that the random device assigns question A to the interviewee and E is the event that the interviewee answers yes to the assigned question. $P_A(E)$ is the probability that an interviewee to whom question A has been assigned answers yes. This is precisely the probability we want to estimate, and for greater convenience we will write Π_A in place of $P_A(E)$. $P_B(E)$ is the probability that an interviewee to whom question B has been assigned answers yes. In our specific example, $P_B(E) = .5$ and $P(A) = .6$ so that

$$\Pi = .6\Pi_A + .4 \times .5 = .6\Pi_A + .2$$

or

$$\Pi_A = \frac{\Pi - .2}{.6}.$$

This relationship between Π_A and Π allows us to convert an estimate of Π into an estimate of Π_A. For example, if 46 interviewees answer yes and 54 answer no in a sample of 100, our estimate of Π is $^{46}/_{100} = .46$ and therefore our estimate of Π_A is $(.46 - .20)/.60 = {}^{26}/_{60} = .43$.

We can find a confidence interval for Π_A by finding a confidence interval for Π and converting this interval into an interval for Π_A. Since 46 out of 100 interviewees answered yes in our example, according to (6.8), a confidence interval for Π with confidence coefficient approximately .95 is given by

$$.46 - .10 \leq \Pi \leq .46 + .10$$

or

$$.36 \leq \Pi \leq .56.$$

Solving for Π_A, we find

$$\frac{.36 - .20}{.60} \leq \Pi_A \leq \frac{.56 - .20}{.60}$$

or

$$.27 \leq \Pi_A \leq .60.$$

PROBLEMS

1 You have observed the results of 100 trials in an attempt to find a point estimate of the success probability p. You then decide that you really want an estimate that is three times as accurate as your present estimate. How many additional trials should you perform to attain this degree of accuracy?

2 You want a point estimate of the success probability p that (with probability .95) does not deviate from the true value p by more than .025. How many trials should you perform to obtain such an estimate?

3 If in Problem 2 you suspect that p is at least .8, how many trials would you perform?

4 You have observed 75 successes in 225 trials.

a. What is your point estimate of the true success probability p?
b. What can you say about the probability that an estimate based on 225 observations does not deviate from the true value p by more than .10; by more than .03?

5 If you observe the results of 400 binomial trials, what can you say about the probability that the corresponding point estimate of the success probability p does not deviate from the true success probability by more than

a. .20;
b. .10;
c. .05;
d. .03;
e. .01;
f. .001?

6 A statistician has told you that a confidence interval of length .10 for the success probability p requires 400 trials. You can perform only 100 trials. How long is your confidence interval going to be? What decision is implicit in your answer?

7 Find confidence intervals for the success probability p corresponding to the following information:

a. $n = 100$, $k = 60$, $\gamma = .95$;
b. $n = 100$, $k = 10$, $\gamma = .95$;
c. $n = 100$, $k = 40$, $\gamma = .95$;
d. $n = 100$, $k = 40$, $\gamma = .90$;
e. $n = 100$, $k = 40$, $\gamma = .99$;
f. $n = 400$, $k = 80$, $\gamma = .99$;
g. $n = 25$, $k = 10$, $\gamma = .90$.

8 If you have observed 60 successes in 100 trials, find a confidence interval ($\gamma = .95$) for the failure probability q.

9 What is the length of a conservative confidence interval ($\gamma = .95$) for the success probability p, if the number of trials equals

a. 50;
b. 100;
c. 200;
d. 400?

10 What is the length of a conservative confidence interval ($\gamma = .95$) for the failure probability q if you observe the results of 100 trials?

11 The confidence intervals for $n = 10$ in Table 6.11 are based on exact computations (beyond the scope of this course). To show that the normal approximation may or may not give satisfactory results for n as small as 10, find 95 percent confidence intervals using Formula (6.9) when

a. $k = 5$;
b. $k = 8$.

12 In Problem 1 of Chapter 2 you were asked to roll a die 120 times. Define success as "the die shows 5 or 6." Find a confidence interval for the success probability p. Does your interval contain the value $p = 1/3$?

13 Problem 2 of Chapter 2 was concerned with 180 imaginary rolls of two dice. Define success as "the two dice show a total of 10 or more." Find a confidence interval for the success probability p. Does the interval contain the value $p = 1/6$?

14 Repeat Problem 13 using the actual data called for in Problem 3, Chapter 2.

15 Consider the data in Table 3.21. Define success as "the second digit equals the first digit." Find a confidence interval for the success probability p. Does the confidence interval contain the value $1/3$?

16 Repeat Problem 15 using the data in Table 3.22.

17 Group Exercise: Take 50 samples of 100 random digits each. For each sample, count the number k of odd digits and find a confidence interval with confidence coefficient .90 using Formula (6.8). How many of your intervals contain the correct success probability $p = 1/2$? It is instructive to represent your results graphically. For this purpose each interval is marked off on a separate horizontal line as follows:

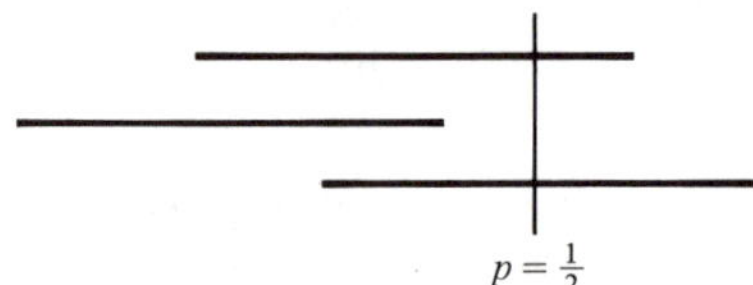

In the present example we happen to know that the true success probability is $1/2$. Thus we know which intervals cover the true success probability and which intervals do not. But, in general, the true success probability p is unknown. We then do not know where p is in relation to the computed confidence limits. All we know is that the long-run frequency with which confidence intervals cover the true parameter value equals the chosen confidence coefficient.

18 An automobile insurance company selected a sample of 500 policies written on subcompact cars. Of the 500 policyholders, 48 had filed claims against the company during the preceding year. Estimate the claim frequency for each 100 subcompact cars insured by the company.

19 A testing program is to be designed to determine the proportion of diabetics in a population. How large a sample is needed to obtain an estimate within one percentage point of the true value of p, if p is known to be less than 10 percent?

20 A student newspaper predicts that 60 percent of the student body is going to vote for candidate X in the upcoming student elections. If the prediction is based on a random sample of 50 students, what are "reasonable" margins of error for this prediction?

21 A newspaper reporter surveyed 200 cars parked in the central district of a city and found that 88 of the cars were parked illegally.

a. Find a point estimate of the probability p that a car is parked illegally.
b. Find a confidence interval with confidence coefficient .90.

22 An analysis of 200 championship chess games showed that 67 of the games were won by white, 43 of the games were won by black. The remaining games resulted in draws. Find a confidence interval ($\gamma = .95$) for the proportion of drawn games in championship play.

23 In a nationwide survey of 2000 families, 400 families indicated that one or more family members participated in the sport of tennis. Find a confidence interval ($\gamma = .99$) for the proportion of families who participate in tennis.

24 In a study of car deaths involving 400 drivers, investigators found evidence of drugs in 122 of the victims. Find a confidence interval ($\gamma = .98$) for the proportion of fatalities involving the use of drugs.

25 160 volunteers participated in an experiment to test a new influenza drug. Even though exposed to influenza virus, 86 of the volunteers did not contract flu. Find a confidence interval ($\gamma = .90$) for the proportion of persons that are protected against flu when using the new drug.

26* The randomized response method is used to ascertain the proportion of female students at a college who use birth control pills. Interviewees are told to toss a coin and answer question *A* if the coin falls heads; question *B*, if it falls tails. These are the questions:

A: Do you use birth control pills?
B: Is the last digit of your student ID number odd?

Of 100 interviewees, 40 answer yes.

a. Find an estimate for the proportion of female students at the college on birth control pills.
b. Find a confidence interval ($\gamma = .90$) for the same parameter.

TESTS OF HYPOTHESES

Basic Ideas

49 In the preceding chapter we discussed estimation of the parameter p of a binomial distribution. Now we take up hypothesis testing. We may have reason to think that the success probability p associated with an experiment like the mental random-digit guesses in Section 19 has some definite value like $p = 1/3$. We should like to know whether our assessment of the value of p can be supported by experimental evidence. In statistical terminology, we want to *test the hypothesis* that p equals $1/3$.

A *statistical hypothesis* is a statement about the distribution of observable quantities. A *test* of a statistical hypothesis is a procedure that tells us whether to *accept* or *reject* the hypothesis under consideration on the basis of experimental information. Our interest in developing procedures for testing statistical hypotheses stems from the fact that our decision to accept or to reject a given hypothesis usually has an important bearing on our future course of action.

Mathematicians on occasion disprove the validity of a mathematical proposition by exhibiting a single counterexample. Statisticians use a similar technique to refute a statistical hypothesis. An example will clarify ideas and, at the same time, call attention to the essential difference between the mathematical and the statistical approaches.

Suppose that we watch a person flip a coin ten times, and on each toss the coin falls heads. We are very likely to think that some cheating is involved. Let us try to establish a parallel between this train of thought and the use of a counterexample to disprove a mathematical proposition:

We set up a

mathematical proposition (example: All prime numbers are odd.)	statistical hypothesis (example: Coin tosses are fair.)

We then consider

a counterexample (2 is both prime and even.)	sample results (10 heads in 10 tosses.)

We conclude that

the mathematical proposition is false.	there are reasons to doubt the correctness of the hypothesis.

As far as the mathematical proposition is concerned, it is false beyond any doubt. As far as the statistical hypothesis is concerned, we cannot say definitely that it is false. Ten heads in ten tosses look very suspicious, but such a result is not impossible under conditions of fair-coin tosses. Fair-coin tosses are characterized by probabilities of $\frac{1}{2}$ for both heads and tails, and the binomial distribution with $n = 10$ and $p = \frac{1}{2}$ says that ten successes in ten trials with success probability $\frac{1}{2}$ occur with probability $(\frac{1}{2})^{10} = \frac{1}{1024}$. But this probability is so small—it is less than 1 in 1000—that we prefer a more plausible explanation than chance. This more plausible explanation is cheating.

More generally, how can we test the hypothesis that the success probability p associated with an experiment has the value p_0? Statisticians often indicate a hypothetical parameter value by the subscript zero. Symbolically, we write

$$H: \quad p = p_0$$

and refer to H as the null hypothesis. (We prefer the symbol H to the more traditional symbol H_0 to indicate a hypothesis that is being tested.)

The coin-tossing example suggests that we base the test of H on the number of successes k that we observe in an experiment involving n independent trials. If H is correct, we expect k to be in the neighborhood of np_0. Thus a value of k close to np_0 does not raise doubts about the correctness of the hypothesis. On the other hand, the more k deviates from np_0, the more untenable the null hypothesis becomes. Where do we draw the dividing line between acceptance and rejection of H? Before we answer this question, we need to introduce some further terminology.

50 A value k that leads to rejection of H is called a *rejection value.* The rejection values form the *rejection* or *critical region.* To determine a test procedure for a statistical hypothesis means to determine a critical region. Whenever our experiment results in a value k in the critical region, we reject the hypothesis being tested; if the value k is not in the critical region, we do not reject the hypothesis. When constructing a critical region, we will always use the following intuitively reasonable principle: Suppose that a critical region contains a value k. Then if k is greater than np_0, the critical region also contains the values $k + 1$, $k + 2, \ldots, n$. If k is less than np_0, the critical region also contains the values $k - 1, k - 2, \ldots, 0$. Thus there are just three possible types of critical regions: critical regions containing only sufficiently large values of k, critical regions containing only sufficiently small values of k, and critical regions containing both sufficiently large and sufficiently small values of k. The three types of critical regions are illustrated as follows:

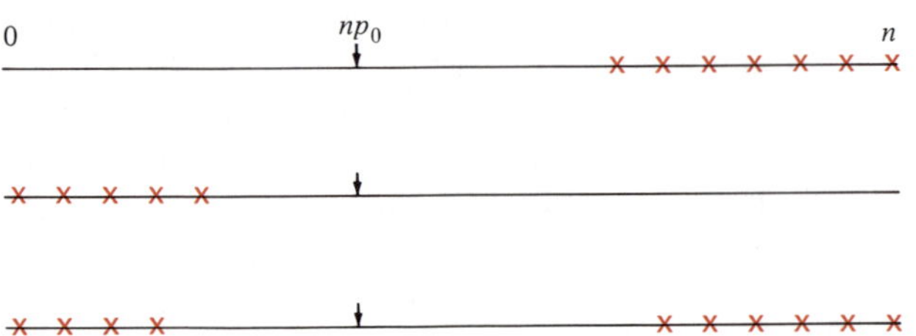

The colored crosses indicate rejection values. We will refer to the first two critical regions as *right-* and *left-tailed,* or simply *one-tailed,* and to the third critical region as *two-tailed.*

How do we choose among these three types and how do we select cutoff points for each particular type? Much of the remainder of this chapter deals with the resolution of these two questions.

The reason why we may hesitate to declare that ten heads in ten coin tosses are an indication of unfair coin tosses is the fact that it is possible for ten fair coin tosses to produce ten heads. We could be accusing an honest person of cheating. Similarly, in our general procedure even if H is true, there is always the possibility that the number of successes resulting from an experiment corresponds to one of the values in the critical region. We would then decide to reject the hypothesis being tested, and this rejection would be incorrect. Clearly, we should like this to happen only rarely. The desire to avoid false rejection suggests the following criterion for selecting a critical region. If the hypothesis is true, the probability of observing a critical number of successes should be small. This probability of falsely rejecting the hypothesis being tested when it is actually true is called the *significance level* of the test and is denoted by the Greek letter α.

The choice of a specific value for α is not strictly a statistical problem. Considerations of the possible consequences of a false rejection—in our case, the unwarranted accusation of cheating—should have considerable influence on the choice of α. Practicing statisticians often use routine levels like .05 or .01, or levels as close to these as possible, but such levels are not prescribed by theory and often simply reflect a desire on the part of the statistician to do as others do without asking why. From the practical point of view, use of a .05-level implies that in approximately 1 out of every 20 cases in which the hypothesis being tested is actually correct we decide to reject it.

Rejecting a true hypothesis is only one of two possible errors associated with hypothesis testing. We may also fail to reject a false hypothesis. Thus if the observed number of successes k does not fall in the critical region, we do not reject the null hypothesis. It is quite common to say under such circumstances that the hypothesis is *accepted.* However, "acceptance" does not imply that the null hypothesis is correct. "Acceptance" means only that there is insufficient experimental evidence for rejection at the chosen significance level.

Often a statistical hypothesis is set up for the purpose of disproving the hypothesis. In the coin-tossing example we may have reason to suspect the honesty of the coin tosser. To catch the tosser in the act of cheating, we set up the hypothesis $p = \frac{1}{2}$ indicative of fair coin tosses and then hope to accumulate sufficient experimental evidence to reject this hypothesis. We then want to know the probability that a test procedure will fail to reject a false hypothesis. We will see that the choice of α has an important bearing on this probability. The smaller a value of α we choose, the smaller the probability of rejecting a false hypothesis becomes.

The two types of error—rejecting an hypothesis that is correct and not rejecting an hypothesis that is false—are basic to hypothesis testing. A second example will show the interrelationship between the probabilities associated with the two types of error. Later we will return to the coin-tossing example.

51 Mrs. X claims that she has extrasensory perception (ESP). Mr. Y doubts her claim and offers her a reward if she can convince him that she has ESP. An experiment is arranged. Mr. Y is going to use 26 cards, a red and a black ace, a red and a black king, and so on, down to a red and a black two. He is going to take one pair at a time and hold one card in his left hand and the other card in his right hand in such a way that Mrs. X can see only the backs of the cards. Nevertheless, she is going to tell which card is red and which is black. The experiment is performed, and Mrs. X makes eight correct and five incorrect calls. Satisfied with her performance, she demands her reward. But Mr. Y does not agree. He contends that Mrs. X has been guessing and that her results can easily be explained in terms of chance. They decide that the problem calls for statistical analysis, which is where we come in.

The problem most certainly calls for statistical analysis, but it would have been much better if a statistician had been consulted before, rather than after, the experiment. As we will see, the whole experiment is most inconclusive. Mrs. X and Mr. Y are left pretty much where they started—Mrs. X convinced of her supernatural powers and Mr. Y as skeptical as ever. However, we are getting ahead of our analysis.

Let p be the probability that Mrs. X indicates the correct arrangement for a single pair of cards. We have to decide whether p is equal to $1/2$ or is greater than $1/2$. The value $p = 1/2$ represents guessing pure and simple, while a value of p greater than $1/2$ indicates something in addition to guessing, whatever it might be. Investigations of extrasensory perception experiments have sometimes brought to light incredibly poor experimental controls.

Finding a Critical Region

52 We have seen that our extrasensory perception problem suggests a test of the hypothesis $p = 1/2$ based on $n = 13$ trials. We should like to reject the hypothesis in question if in reality p is greater than $1/2$. Using statistical terminology, we want to test the hypothesis

$$H: \quad p = \frac{1}{2}$$

against the *alternative*

$$A: \quad p > \frac{1}{2}.$$

If our hypothesis is correct, that is, if Mrs. X does not have extrasensory perception, we would expect her to make in the neighborhood of $np_0 = 13 \times 1/2 = 6.5$ correct identifications simply by guessing the order of each of the 13 pairs of cards. Only if Mrs. X makes a substantially larger number of correct identifications, should we concede that Mrs. X is not guessing and allow her to receive the promised reward.

Let k stand for the number of correct identifications made by Mrs. X. Our discussion implies that an appropriate critical region for testing H consists of sufficiently large values of k. Had the alternative stated that $p < 1/2$, a left-tailed critical region would have been appropriate. Once we have determined the type of critical region, our next problem is to decide how large a value of k is sufficiently large for purposes of rejection. We look at various possible critical regions and compute the corresponding significance levels α. If we are willing to concede that Mrs. X has extrasensory perception if she identifies all 13 pairs correctly, the critical region consists of the single value $k = 13$. The significance level associated with this critical region is $\alpha = b(13)$, where $b(k)$ is the probability of observing k successes in 13 trials with success probability $p = 1/2$. According to Table A, we find that α is zero to three decimal places or $\alpha < .0005$. Few people would insist on a critical region with such a small significance level. If we are willing to concede extrasensory perception provided that Mrs. X makes at least 11 correct identifications, we use a critical region containing the values $k = (11, 12, 13)$. This critical region has significance level

$$\alpha = b(11) + b(12) + b(13) = .010 + .002 + .000 = .012,$$

still rather small. Table 7.1 gives significance levels for all critical regions of the type $k \geq r$, where r is at least 7.

TABLE 7.1 **Significance Levels**

r	7	8	9	10	11	12	13
Significance Level α *for Critical* Region k $\geq r$	.500	.291	.134	.047	.012	.002	.000

If we have our heart set on a test with significance level approximately .05, we have to use the critical region containing the values $k = 10$, 11, 12, and 13. Since Mrs. X identified only 8 out of 13 pairs, there is no reason to reject the hypothesis $p = \frac{1}{2}$. Even if we should have been willing to use a significance level as large as .13, the available information still does not suggest rejection of the null hypothesis. This is why Mr. Y was unimpressed by the performance of Mrs. X and why he felt justified in withholding the reward.

However, there is another side to the picture. Let us assume, just for the sake of argument, that Mrs. X is not guessing. What are her chances of collecting the promised reward? To have something more specific to talk about, let us assume that she has probability $p = \frac{2}{3}$ of making a correct identification. Such a value of p does not necessarily imply clairvoyant powers. Mr. Y may have a tendency to alternate red and black according to a regular pattern. If Mrs. X learns from observation, she should be able to achieve a rather high score. Mr. Y can prevent such a possibility by using a mechanical randomizing procedure—like the toss of a coin or the use of random digits—to determine for each pair of cards which cards goes on the right and which on the left. But, at this moment, we are not concerned with this aspect of the matter. We only want to know the probability that Mrs. X makes a number of correct identifications sufficiently high to reject the hypothesis of guessing. Now we need probabilities based on $p = \frac{2}{3}$ rather than $p = \frac{1}{2}$. These probabilities are given in the middle row of Table 7.2 labeled *rejection* probability and are denoted by $1 - \beta$.

If we use the test with significance level approximately .05, Mrs. X has probability .332, less than one chance in three, of showing that she is not guessing. This probability goes down still further if we use a more stringent critical region.

The significance of these probabilities will perhaps become clearer if we give a reinterpretation of our problem. Suppose that, at the end of

TABLE 7.2 **Test Probabilities**

r	7	8	9	10	11	12	13
Significance Level α *for Critical Region* $k \geq r$	.500	.291	.134	.047	.012	.002	.000
Rejection Probability $1 - \beta$ *when* $p = 2/3$	.896	.759	.552	.322	.139	.039	.005
Failure Probability β *when* $p = 2/3$	.104	.241	.448	.678	.861	.961	.995

a course, an instructor announces that every student will have to take an examination consisting of 13 true-and-false questions. To get a passing grade for the course, a student would have to answer at least 10 questions correctly. Otherwise the student would fail. Those are exactly the requirements Mrs. X has to meet to collect the reward, if we are to apply a test with significance level .05.

Of course, students who know the course material well should have no difficulty answering at least 10 out of 13 questions correctly. But what about a marginal student? To be more specific, let us take a student who knows the answers to approximately $2/3$ of the type of questions that are customarily asked on an examination of this type. Such a student would seem to be entitled to at least a low pass. But what does our table tell us? Such a student has less than one chance in three of getting a passing grade for the course. Perhaps the issue becomes even clearer when we look at the last line in Table 7.2. There we find the probabilities β that our marginal student fails the course. Alternatively, these are the probabilities that Mrs. X is unable to convince Mr. Y of her special abilities.

Type 1 and Type 2 Errors

53 Before continuing the discussion of hypothesis testing, we pause to review what we have learned so far. A test of a statistical hypothesis can go wrong in two completely different ways depending on whether the hypothesis happens to be true or false. When the hypothesis is true and, on the basis of sample evidence, we decide to reject it, we commit what is called a *type 1 error.* When the hypothesis is false and, on the basis of sample evidence, we do not decide to reject it, we commit what is called a *type 2 error.* The possibilities that may occur when testing a hypothesis are represented schematically in Table 7.3.

In the extrasensory perception example, a type 1 error occurs if Mrs. X is guessing but happens to make a sufficiently large number of correct guesses to convince Mr. Y of her (nonexistent) powers of ESP. The probability of this happening is α, the significance level chosen for the test. Should a type 1 error occur, Mrs. X receives an undeserved reward.

TABLE 7.3 **Correct and Incorrect Decisions in Hypothesis Testing**

Action Taken as Result of Test	*Unknown True Situation* — *Hypothesis Is True.* ($p = \frac{1}{2}$)	*Unknown True Situation* — *Alternative Is True.* ($p = \frac{2}{3}$)
Accept Hypothesis: (Give no Reward.)	Correct Decision Probability: $1 - \alpha$	Type 2 Error Probability: β
Reject Hypothesis. (Give Reward.)	Type 1 Error Probability: α	Correct Decision Probability: $1 - \beta$

A type 2 error occurs if one way or another Mrs. X's probability of making a correct identification is greater than $\frac{1}{2}$ ($\frac{2}{3}$, in our discussion), but nevertheless fails to identify correctly a sufficiently large number of pairs of cards to satisfy Mr. Y. If a type 2 error occurs, Mrs. X is denied a deserved reward.

When we set up a critical region for testing a hypothesis and insist on a small significance level, we think only in terms of the probability of committing a type 1 error. Table 7.2 shows that the probabilities of type 1 and type 2 errors are inversely related. As we try to decrease one probability, we increase the other, at least as long as we keep other aspects of the experiment unchanged.

Clearly, we cannot have both probabilities arbitrarily small at the same time in the extrasensory perception experiment. We have to compromise somewhere. It often happens that one error is much more serious than the other. In such a case it is reasonable to make the probability of the more serious error smaller than that of the less serious error. But in our example it seems reasonable to make the two error probabilities as similar as possible. This occurs for the critical region $k \geq 8$. Of course, neither error probability is very satisfactory. The probability of giving an unearned reward is .29 and the probability of withholding an earned reward is .24.

The only real way out of the difficulty is to change the experimental setup. Thirteen observations are simply not enough to discriminate satisfactorily between the two probabilities $\frac{1}{2}$ and $\frac{2}{3}$. This is exactly what a statistician could have pointed out before the start of the whole experiment. What is more, a statistician could have determined how many trials should have been performed to reduce both probabilities to reasonable levels.

54 To see how the error probabilities decrease as the number of observations increase, let us look at an example involving 50 observations. In particular, let us think of a true-and-false examination containing 50 questions.

This time the parameter p is the probability that a student answers a question correctly. Since we deal only with true-and-false questions, even a completely unprepared student can achieve a value of $p = \frac{1}{2}$ by tossing a coin and, say, answering T (for true) every time the coin falls heads and F (for false) when it falls tails. We want to answer two questions. What is the probability that a totally unprepared student receives a passing grade? What is the probability that a prepared student receives a failing grade?

We interpret "totally unprepared" to mean that a student guesses whether to mark a given statement true or false by using either real or imagined coin tosses. In terms of the parameter p, an unprepared student is one for whom $p = \frac{1}{2}$. A prepared student is one for whom p is greater than $\frac{1}{2}$. More specifically, we will assume, at least for the moment, that $p = \frac{2}{3}$, indicating a fairly marginal student. Our procedure can then be viewed as a test of the hypothesis

$$H: \quad p = \frac{1}{2}$$

against the alternative

$$A: \quad p = \frac{2}{3}$$

using 50 observations as the basis of our test.

Before we compute the probabilities of type 1 and type 2 errors, it is helpful to state the specific nature of these errors. We commit a type 1 error when we reject a true null hypothesis. In our example a type 1 error is committed when the instructor gives a passing grade to a student who deserves to be failed. A type 2 error occurs if the null hypothesis is accepted when in reality the alternative hypothesis happens to be true. In our example a type 2 error is committed when on the basis of his examination score our student receives a failing grade, even though his general knowledge warrants a passing grade. We mentioned earlier that the seriousness of the given errors should determine to some extent the α and β probabilities that we are willing to tolerate. The present example shows that different persons may have different ideas about the relative seriousness of the two errors. A student taking the examination would hardly worry about the type 1 error, receiving a passing grade, when a failing grade should have been received. But the student would certainly object to receiving a failing grade when a passing grade should have been awarded. The instructor, on the other hand, may feel that both kinds of errors are equally undesirable.

We will now compute actual probabilities. Let k be the number of questions answered correctly. A student who guesses can expect to answer approximately 25 questions correctly by chance alone. A student who has some knowledge of the subject matter should be able to do better. Thus sufficiently large values of k indicate that the null hypothesis should be rejected. The critical region for our test consists of sufficiently large values of k, that is, all values of k that are greater than or equal to some value r.

Table 7.4 gives both α and β probabilities for various choices of r. We note that the error probabilities are considerably lower than those in Table 7.2. But error probabilities of .10 and .13 corresponding to $r = 30$ are still somewhat high, suggesting that even 50 questions on a true-and-false examination is not sufficiently discriminatory.

TABLE 7.4 **Type 1 and Type 2 Error Probabilities**
H: $p = 1/2$, A: $p = 2/3$, $n = 50$

r	29	30	31	32
$\alpha = P(k \geq r)$ when $p = 1/2$	.161	.101	.059	.032
$\beta = P(k < r)$ when $p = 2/3$	.076	.126	.196	.287

The Power Curve

55 Some readers may have wondered why we have concentrated on the value $p = 2/3$ in discussing type 2 errors. There is really no particular reason. Any other value of p greater than $1/2$ could have served equally well. Indeed, when evaluating the properties of a test procedure, statisticians like to look at several different p-values. Thus, in evaluating the test that rejects the null hypothesis $p = 1/2$, if the observed number of successes is at least 30 ($r = 30$), we should consider the kind of information contained in Table 7.5. Such information can be plotted on a graph by marking p along the horizontal axis and the rejection probability along the vertical axis as in Figure 7.6. The resulting curve is called the *power curve* of the test. Power curves enable the statistician to evaluate the effectiveness of a given test procedure and to compare two or more competing test procedures.

TABLE 7.5 **Rejection Probabilities**
$p_0 = 1/2$, $n = 50$, Critical Region: $k \geq 30$

True p-value	.45	.50	.55	.60	.65	.70
Rejection Probability	.024	.101(=α)	.286	.561	.814	.952

In the present problem, where we want to decide whether or not a student is to receive a passing grade, it has little meaning to ask what happens when $p = .45$ or some other value smaller than $1/2$. One possible interpretation of such a p-value is that the student is not guessing but has actually acquired some incorrect information leading to a reduction in overall performance. Table 7.5 reveals that a student whose p-value is .45 has probability .024 of receiving a passing grade. Information like this is important in situations that are mathematically

FIGURE 7.6

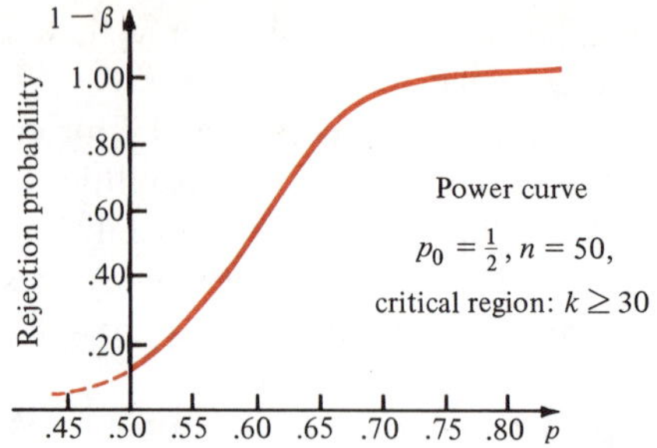

identical to the present problem but that involve different practical circumstances. We consider the following (oversimplified) example.

56 A standard medical treatment has been observed to produce undesirable side effects in 50 percent of the patients to whom it is given. It is claimed that a modification of the treatment, which is considerably more complicated than the original treatment, would decrease this percentage without changing the effectiveness of the treatment itself. The question that has to be answered is whether doctors should use the modified treatment in the future in place of the old treatment.

Here we have a situation where it is necessary to choose between two possible courses of action, recommending or not recommending the use of the modified treatment. This is typical of the situation that calls for an appropriate test of a statistical hypothesis, where we decide on one course of action if the test should lead to acceptance of the hypothesis being tested and a different kind of action if the test should reject the hypothesis. In the present situation, since the modified treatment is still experimental, it is not known what percentage of patients to whom the modified treatment is given will stay free of side effects. Let us denote this unknown proportion by p. (With this notation, a success is a patient who undergoes the modified treatment and does not develop any side effects.) If p should be $\frac{1}{2}$ or less, use of the modified treatment in place of the old treatment is highly undesirable. Not only is the modified treatment more complicated, but it also represents no improvement over the old treatment and actually may be worse. On the other hand, the more p exceeds $\frac{1}{2}$, the more desirable it becomes to replace the old treatment by the modified treatment. The problem of choosing between the two possible courses of action can be decided statistically by setting up a test of the hypothesis $p = \frac{1}{2}$ and choosing a critical region that contains only large k-values. Acceptance of the hypothesis $p = \frac{1}{2}$ leads to the recommendation that the modified treatment should not be used in the future; rejection leads to the opposite recommendation. If, in particular, we carry out an experiment involving 50 patients and decide to recommend use of the modified treatment in the event that at least 30 of these 50 patients do not show any side effects, then the information in Table 7.5 (and Figure 7.6) allows the following reinterpretation. If $p = \frac{1}{2}$, the same as for the old treatment, the probability is .101, or 1 chance in 10, that we are going to recommend use of the more complicated modified treatment even though the modification does not change the incidence of side effects. There is 1 chance in 40 (probability .024 to be exact) that we are going to recommend the treatment modification even though only 45 percent of all patients are going to stay free of side effects. On the other hand, if p should equal .70, there are about 19 chances in 20 (probability .952 to be exact) that we are going to recommend introduction of the modified treatment. With such information at their disposal, medical authorities can then decide whether an experiment involving 50 patients provides sufficient information to make the kind of decision that is required or whether possibly a larger (or smaller) experiment is in order.

Sample Size

57 The problem of determining an appropriate sample size is often solved as follows. In addition to the hypothetical value p_0 and the associated type 1 error probability α, we select another value p_1 such that, if the true success probability is p_1, we should like the probability of a type 2 error to be some small quantity β. In other words, p_0 and p_1 are selected in such a way that we are very eager to do one thing (stay with the old treatment in our medical example) if $p = p_0$ or less and do something else (switch to the modified treatment) if $p = p_1$ or larger. Any test of the type discussed in this chapter that satisfies the given two requirements has a power curve like the one illustrated in Figure 7.7. If $p \leq p_0$, the probability of deciding to act as if $p = p_1$ is less than or equal to α. If $p \geq p_1$, the probability of deciding to act as if $p = p_0$ is less than or equal to β. It can then be shown that the number n of observations required for such a test is approximately

$$n \doteq \frac{(z_0\sqrt{p_0q_0} + z_1\sqrt{p_1q_1})^2}{(p_1 - p_0)^2},$$

FIGURE 7.7 Power Function and Sample Size Determination

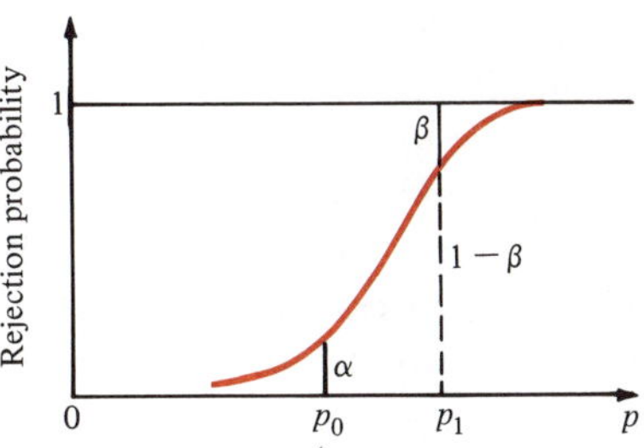

where z_0 is read from Table C corresponding to $\alpha' = \alpha$ and z_1 is read from Table C corresponding to $\alpha' = \beta$.

EXAMPLE 7.8 If $p_0 = \frac{1}{2}$, $\alpha = .10$, $p_1 = .7$, $\beta = .05$, then

$$n = \frac{(1.282\sqrt{\frac{1}{2} \times \frac{1}{2}} + 1.645\sqrt{.7 \times .3})^2}{(.7 - .5)^2} = 49.$$

(Note that the test underlying Table 7.4 has $p_0 = \frac{1}{2}$, $\alpha = .101$, $p_1 = .7$, $\beta = 1 - .952 = .048$, and $n = 50$, in very close agreement with our present result.)

In practice, experimenters often like to have tests that are able to discriminate with small error probabilities between values p_0 and p_1 that differ very little. Such discrimination is possible only on the basis of a large number of observations. If the experimenter does not have the facilities to conduct such a large-scale experiment, a decision must be made as to whether to scale down the requirements or to abandon the experiment altogether. It is certainly better to be forewarned than to find out after performing the experiment that the information obtained is insufficient to make a reliable decision, as in the experiment conducted to prove or disprove Mrs. X's powers of clairvoyance.

Two-sided Alternatives

58 We now return to the coin-tossing example. We are interested in finding out whether certain coin tosses are fair or unfair. For fair coin tosses the probability of heads is $\frac{1}{2}$. Thus we set up the hypothesis

$$H: \quad p = \frac{1}{2}.$$

As in Section 49 we will assume that we have observed the results of $n = 10$ tosses. Which outcomes of these coin tosses will lead us to conclude that cheating has been involved? In statistical terms, if k denotes the observed number of heads, which values of k should be in the critical region leading to the rejection of H?

This looks like the corresponding problem in the extrasensory perception controversy between Mrs. X and Mr. Y. In both cases the hypothesis to be tested states that the true success probability is $\frac{1}{2}$. Why then are we considering this problem all over again? After all, conducting 10 trials rather than 13 does not represent such an important difference; this is perfectly true. But the two examples differ in another much more important aspect. In the extrasensory perception example, the critical region consisted of large values of k indicating that the true success probability p was presumably greater than $\frac{1}{2}$. In the present example, a cheater who knows how to produce too many heads presumably also knows how to produce too few. This time we want to reject the null hypothesis if the true success probability is either greater than $\frac{1}{2}$ or less than $\frac{1}{2}$. Thus we want to test the hypothesis $p = \frac{1}{2}$ against the alternative

$$A: \quad p \neq \frac{1}{2}.$$

Such an alternative is said to be *two-sided* because it implies that we may have $p < \frac{1}{2}$ or $p > \frac{1}{2}$. The appropriate critical region is *two-tailed*, containing both small and large values of k. Let us see what critical regions are available and what the associated significance levels are according to Table A:

$$\begin{array}{ll} \text{(i) } k = (0 \text{ or } 10) & \alpha = .002 \\ \text{(ii) } k = (0, 1, 9, 10) & \alpha = .022 \\ \text{(iii) } k = (0, 1, 2, 8, 9, 10) & \alpha = .110 \end{array}$$

The α-levels follow from the binomial probability table for $n = 10$ and $p = \frac{1}{2}$. Thus for region (ii), $\alpha = b(0) + b(1) + b(9) + b(10) = .022$. If we decide to reject the hypothesis $p = \frac{1}{2}$ whenever we observe 0, 1, 9, or 10 heads in 10 tosses of the coin, we have roughly 2 chances in 100 of declaring the coin tosses unfair when in reality $p = \frac{1}{2}$.

Although many statisticians would feel that a significance level of .110 is too large, it is better not to take such a rigid attitude. Our example shows one thing very clearly. With ten observations we cannot hope to recognize anything but the crudest violations of the null hypothesis. For example, the second critical region requires at least a nine-to-one split between heads and tails before we can reject the null hypothesis. Such a result is not very likely to occur unless the true probability of heads is very close to zero or very close to one. Thus we have practically no chance of finding out that the coin tosses are not fair unless the bias is very strong.

The situation gets progressively better as we take more and more observations. Table 7.9 lists critical regions that have α-levels as close to .05 as we can get. Thus for $n = 10000$, a split of at least 49 to 51 among heads and tails assures rejection of the null hypothesis. Such a split is very likely to occur when the coin is only slightly biased in favor of heads or tails.

TABLE 7.9 Critical Regions for Testing H: $p = \frac{1}{2}$ with Significance Level $\alpha = .05$

n	*Critical Regions*	*Split*
100	(0, . . . , 40, 60, . . . , 100)	40:60
1000	(0, . . . , 469, 531, . . . , 1000)	47:53
10000	(0, . . . , 4901, 5099, . . . , 10000)	49:51

59 When testing the hypothesis $p = \frac{1}{2}$, symmetry considerations suggest that we put the value $k = 0$ in the critical region along with $k = n$, the value $k = 1$ along with $k = n - 1$, and so on. But suppose that we want to test the hypothesis $p = .4$ on the basis of 10 observations. If the hypothesis is correct, we expect to observe in the neighborhood of 4 successes. We should decide to reject the hypothesis only if the number of successes actually observed differs sufficiently from 4. However, this time earlier symmetry considerations do not apply since $b(0)$ does not equal $b(10)$, $b(1)$ does not equal $b(9)$, and so on. In such a case it is usually appropriate to build up the two tails of the critical region subject to the following two conditions:

(i) The probability of all k-values in the critical region should correspond to the chosen significance level α.
(ii) The two tails should as nearly as possible have equal probabilities.

In the example of testing the hypothesis $p = .4$, if we want approximately $\alpha = .10$, we can use the critical region $k = (0, 1, 7, 8, 9, 10)$. For this critical region the lower tail has probability $\alpha_1 = b(0) + b(1) = .006 + .040 = .046$, while the upper tail has probability $\alpha_2 = b(7) + b(8) + b(9) + b(10) = .042 + .011 + .002 + .000 = .055$, so that the significance level is $\alpha = \alpha_1 + \alpha_2 = .101$. When the number n of trials is small, it is at times impossible to equalize the two tail probabilities satisfactorily. However, equalization becomes less and less of a problem as n increases.

60 Type 2 error probabilities are computed as usual.

EXAMPLE 7.10 Find the probability of committing a type 2 error for the test in Section 59, when the true success probability equals .2. We commit a type 2 error when we accept the hypothesis being tested ($p = .4$) even though it is false ($p = .2$). In our example the hypothesis is accepted if we observe at least 2 but no more than 6 successes among our 10 trials. Thus the probability of committing a type 2 error equals $\beta = b(2) + \cdots + b(6) = .302 + \cdots + .006 = .623$ using the table of binomial probabilities for $n = 10$ and $p = .2$.

The power of our test against the alternative $p = .2$ equals $1 - \beta = .377 = b(0) + b(1) + b(7) + b(8) + b(9) + b(10)$, the probability that k takes a value in the critical region.

Normal Approximations

61 So far we have assumed that we have a table of the appropriate binomial distribution to set up a critical region based on the observed number of successes k. We know from Chapter 5 that if the number of

trials is large, binomial probabilities can be approximated using the normal distribution. This approximation provides us with a simple procedure for testing the hypothesis $p = p_0$. When this hypothesis is true, the quantity

7.11 $$z = \frac{k - np_0}{\sqrt{np_0q_0}}$$

has approximately a standard normal distribution. We can then replace a critical region based on k by a critical region based on z, where large positive values of z correspond to large values of k and large negative values of z correspond to small values of k. Table C provides the necessary information to carry out tests based on the quantity z.

EXAMPLE 7.12 We have observed 100 coin tosses resulting in 37 heads and 63 tails. Can we reject the hypothesis $p = \frac{1}{2}$ against the alternative $p \neq \frac{1}{2}$ at significance level .05? According to Table 7.9, the answer is yes, since $k = 37$ falls in the lower tail of the appropriate critical region. For the normal approximation, since we are interested in a two-sided test, we find the significance level .05 in the column labeled α'' of Table C and read the corresponding z-value 1.960 in the last column. It follows that our hypothesis should be rejected if the statistic z given by (7.11) is either greater than $+1.96$ or smaller than -1.96 (see Fig. 7.13). For our example,

$$z = \frac{37 - 100 \times \frac{1}{2}}{\sqrt{100 \times \frac{1}{2} \times \frac{1}{2}}} = \frac{37 - 50}{\sqrt{25}} = -\frac{13}{5} = -2.6.$$

FIGURE 7.13

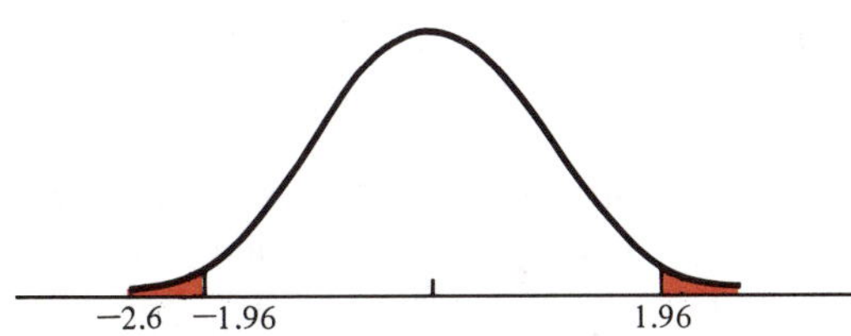

Since -2.6 is smaller than -1.96, we reject the hypothesis of fair coin tosses.

EXAMPLE 7.14 If 400 patients participate in the medical experiment discussed in Section 56 and 221 patients show no side effects, can we conclude that the proportion p of patients who do not show any side effects when using the new treatment is greater than $\frac{1}{2}$? Considering the substantial greater expense of the new treatment, we decide on the significance level .01. Since we are testing the hypothesis $p = p_0 = \frac{1}{2}$ against the one-sided alternative $p > \frac{1}{2}$, we find the significance level .01 in the column labeled α' in Table C and read the corresponding z-value 2.326. At significance level .01, our hypothesis should be rejected provided that the statistic z given by (7.11) is greater than $+2.326$. In our example,

$$z = \frac{221 - 400 \times \frac{1}{2}}{\sqrt{400 \times \frac{1}{2} \times \frac{1}{2}}} = \frac{221 - 200}{\sqrt{100}} = \frac{21}{10} = 2.1.$$

Since $2.1 < 2.326$, we conclude that our evidence is not sufficiently strong to reject the null hypothesis.

When testing the hypothesis $p = p_0$ against the one-sided alternative $p < p_0$ at significance level .01, we should have $z < -2.326$ to reject the null hypothesis.

Summary Remarks

62 Before continuing, it is helpful to summarize certain general ideas. When we test a hypothesis H like $p = p_0$, we are seeking guidance about which of two possible actions to take, an action 1 that is appropriate if $p = p_0$ or an action 2 that is appropriate when the true value of p differs from the hypothetical parameter value p_0 in some way or other. When we set up a test of the hypothesis H, we should keep in mind that acceptance of H has the practical implication of taking action 1; rejection of H has the practical implication of taking action 2. When we talk of type 1 and type 2 errors (or α-errors and β-errors, as they are sometimes called), we are really talking about the error of taking action 2 when we should have taken action 1 and vice versa. This realization throws new light on the meaning of the power curve. Whatever the true value p, the power curve indicates the probability with which we are going to take action 2 when we use the given test procedure.

Since we reject H and take action 2 whenever k falls in the critical region, the critical region should consist of k-values that are indicative of the alternative parameter values for which action 2 is appropriate. It is good practice to state along with the null hypothesis H the alternative A against which the test is to be effective. Thus for the coin problem, we test the null hypothesis H: $p = \frac{1}{2}$ against the alternative A: $p \neq \frac{1}{2}$. This alternative requires a two-tailed critical region. In the medical problem we test the hypothesis H: $p = \frac{1}{2}$ against the one-sided alternative A: $p > \frac{1}{2}$, suggesting an upper-tail critical region. In such a problem it often is more logical to write the null hypothesis as $p \leq \frac{1}{2}$, since we usually are even more interested in taking action 1 when $p < \frac{1}{2}$ than when $p = \frac{1}{2}$. The power curve in Figure 7.6 shows that the critical region consisting of large k-values achieves this purpose: If $p = \frac{1}{2}$, the probability of taking action 1 is $1 - \alpha$, and as p becomes smaller, this probability increases toward 1.

In general, there are three different test situations:

(i) H: $p = p_0$ against A: $p \neq p_0$, requiring a two-tailed critical region;
(ii) H: $p \leq p_0$ against A: $p > p_0$, requiring a critical region consisting of large k-values;
(iii) H: $p \geq p_0$ against A: $p < p_0$, requiring a critical region consisting of small k-values.

Our medical problem could have been formulated as a type (iii) problem by defining success as the occurrence, rather than nonoccurrence, of side effects.

It should be clear from the discussion in this chapter that a knowledge of the power of a test is of great importance in the proper conduct of an experiment. Power computations are relatively simple for the type of problems that we have been discussing in this chapter. However, this is no longer the case for the problems to be discussed in subsequent chapters. While we will still mention the general principles that we have developed in this chapter, we will not try to carry out further power computations. Such computations have to be reserved for more advanced treatments.

PROBLEMS

1 For each of the following hypotheses, set up possible critical regions and determine the associated significance levels. In all cases assume samples of size 20.

a. H: $p = .50$ against A: $p > .50$;
b. H: $p = .50$ against A: $p < .50$;
c. H: $p = .30$ against A: $p > .30$;
d. H: $p = .30$ against A: $p < .30$.

2 You are testing the hypothesis $p = .30$ against the alternative $p < .30$ and have decided to reject the hypothesis if in 25 trials you observe fewer than 5 successes.

a. What is the significance level of your test?
b. Find the probability of committing a type 2 error if the true value of p is .20. Repeat for $p = .10$, $p = .05$.
c. What is the probability of accepting the hypothesis if $p = .40$? By accepting the hypothesis in this case, have you committed an error (if so, which error), or have you made the correct decision?
d. Use the probabilities computed for parts a, b, and c of this question to construct the power curve for the test.

3 You want to test the hypothesis $p = .2$ against the alternative $p > .2$ using 13 trials. Find the appropriate critical region corresponding to a significance level .10.

4 You are testing the hypothesis $p = .4$ on the basis of 25 trials. You decide to reject the hypothesis if either $k \leq 5$ or $k \geq 15$, where k is the observed number of successes.

a. Against what alternative is this test appropriate?
b. What is the significance level of the test?
c. If the true value of p is .2, what is the probability that the test accepts the hypothesis being tested?
d. If the event in c occurs, which error, if any, has been committed?

5 Consider the test that rejects the hypothesis $p = .6$ if in 20 trials at most 9 successes are observed.

a. Against what alternative is the test appropriate?
b. What is the significance level of the test?

6 You are testing the hypothesis $p = .1$ against the alternative $p > .1$ on the basis of 100 trials. As your critical region, you have chosen $k \geq 15$. According to the normal approximation, what is the significance level of this test?

7 Use tables of exact binomial probabilities (see Bibliography) to check the entries in

a. Table 7.4;
b. Table 7.5.

8 a. How many true-and-false questions should be included on an examination so that a student who is guessing has 1 chance in 20 of receiving a passing grade while a student who knows the answers to 60 percent of the type of questions appearing on the examination has 1 chance in 10 of failing?
b. How many of these questions would a student have to answer correctly to obtain a passing grade?
c. What is the probability that a student who knows the answers to 70 percent of the type of questions passes the examination?

9 Find the significance level of the test of the hypothesis $p = \frac{1}{2}$, $n = 10$, where the critical region is given as $k = (0, 1, 2, 3, 7, 8, 9, 10)$.

10 Set up possible critical regions to test each of the following hypotheses when a sample of the indicated size is available. State the significance level of each test. Consider two-sided alternatives.

a. $p_0 = .50$, $n = 20$;
b. $p_0 = .50$, $n = 25$;
c. $p_0 = .60$, $n = 10$;
d. $p_0 = .20$, $n = 20$.

11 You are testing the hypothesis $p = .6$ using 25 observations. You decide to reject the hypothesis if either $k \leq 10$ or $k \geq 20$, where k is the observed number of successes. Find the significance level of your test using

a. exact binomial probabilities;
b. the normal approximation.

12 Verify that the critical regions in Table 7.9 for sample sizes 100, 1000, and 10000 have significance level .05.

13 Find at least 7 points on the power curve of the test of the hypothesis $p = \frac{1}{2}$ using the critical region $k = (0, 1, 2, 8, 9, 10)$. The total number of trials is 10.

14 Past experience shows that, if a certain machine is adjusted properly, 5 percent of the items turned out by the machine are defective. Each day the first 25 items produced by the machine are inspected for defects. If three or fewer defects are found, production is continued without interruption. If four or more items are found to be defective, production is interrupted and an engineer is asked to adjust the machine. After adjustments have been made, production is resumed. This procedure can be viewed as a test of the hypothesis $p = .05$ against the alternative $p > .05$, p being the probability that the machine turns out a defective item. In test terminology, the engineer is asked to make adjustments only when the hypothesis is rejected.

a. What is the critical region of the test?
b. What is the significance level of the test?
c. What is the practical meaning of a type 1 error?
d. What is the practical meaning of a type 2 error?
e. If the machine has gone out of adjustment and produces 20 percent defectives, what is the probability of a type 2 error?
f. The foreman would like to have at most 1 chance in 100 of stopping production needlessly for machine adjustments. What is the smallest number of items out of 25 that have to be found defective to achieve this purpose?

15 The data in Table 12.11 represent yearly incomes of 20 families in a certain community. Each family had been instructed to report the family income to the nearest 100 dollars. Do you think that the respondents heeded the instruction? Justify your answer.

16 Use the data for Problem 6 in Chapter 10 to test the hypothesis that the probability that a person has brown eyes is $\frac{1}{2}$.

17 Use the data for Problem 6 in Chapter 10 to test the hypothesis that the probability that a person has dark hair is $\frac{1}{2}$.

18 In Problem 1 of Chapter 2 you were asked to roll a die 120 times. Define success as "the die shows 5 or 6." Test the hypothesis that $p = \frac{1}{3}$.

19 Problem 2, Chapter 2, was concerned with 180 imaginary rolls of two dice. Define success as "the two dice show a total of 10 or more." Test the hypothesis that $p = \frac{1}{6}$.

20 Repeat Problem 19 using the actual data called for in Problem 3, Chapter 2.

21 According to the 1960 Census, about one family in four had more than three children. Problem 5 of Chapter 1 reports on a recent survey involving the number of children of families living in a certain community. Would you say that the census information is still applicable to this community?

22 Consider Problem 22 of Chapter 6. Test the hypothesis that black and white have equal probability of winning a game of chess. (*Hint:* Ignore drawn games.)

23 In the early days of zip code, a press service conducted the following experiment: 720 letters were mailed in pairs, one letter in each pair bearing zip code, the other not. For the 360 pairs, the letter with zip code arrived before the letter without zip code 101 times. The reverse event occurred 65 times. In all other cases both letters arrived together. What conclusion concerning the benefit of zip code do you draw?

24 National statistics show that approximately 40 percent of the applicants to American medical schools succeed in gaining admission. A well-known private college reports that 43 of its graduates applied to medical schools and 30 were admitted. Is the college justified in claiming that its graduates are doing better than the national average?

25 You are told that, in a certain coin-tossing experiment, the coin fell heads 44 percent of the time and tails 56 percent of the time. At significance level .05, can you reject the hypothesis that coin tosses were fair? (*Hint:* Assume first that the coin was tossed 100 times. Then assume that the coin was tossed 400 times.)

26 A nationwide survey involving 1600 persons contained 730 persons of age ≤ 25. Are these findings in agreement with the widespread belief that the median age of Americans is less than 25 years? (According to a recent report by the Census Bureau, the median age is actually 27.9 years.)

27 20 persons participate in a wine-tasting experiment. Each person is presented with two glasses of wine filled from bottles labeled A and B, respectively. Each person is then asked which wine he/she prefers. Actually both bottles contain the same kind of wine. If at least 15 persons prefer A over B or at least 15 prefer B over A, would you say that the 20 persons participating in the wine-tasting experiment are expert tasters? Why?

28 You want to find out whether fatal automobile accidents are less likely to occur on a weekday than on Saturday or Sunday.

a. Set up an appropriate hypothesis and test it using the data provided in Table 9.1. (*Hint:* Call a fatal accident occurring on a weekend a success, and one occurring on a weekday, a failure.)
b. Suppose that we had information on the number of *individual* fatalities occurring in the 175 accidents. If we want to carry out a similar test using the total number of fatalities, which assumption underlying the methods of the present chapter is likely to be violated?

8

TO TEST OR NOT TO TEST

63 Hypothesis testing as discussed in the previous chapter is appropriate when we are faced with the necessity of choosing one of two possible courses of action on the basis of experimental evidence. Experimenters frequently formulate their problems as problems of hypothesis testing even though experimental data have been collected primarily to gain insight rather than to make definite decisions. In this chapter we will discuss two alternative approaches that are usually more satisfactory in such situations than a test of a hypothesis: (i) replacing the dichotomous accept–reject decision inherent in hypothesis testing by a more graduated statement called the *descriptive level*, or (ii) changing the testing problem to an estimation problem by computing an appropriate confidence interval.

Descriptive Levels

64 By the *descriptive level* (also called the *probability level* or *P-value*), we mean the smallest significance level α at which the observed test result would be declared significant, that is, would be declared indicative of rejection of the hypothesis under consideration.

For binomial experiments, the descriptive level is computed as follows. Suppose that we are interested in the hypothesis $p = p_0$ and have observed k successes in n trials. For alternatives $p > p_0$, the descriptive level is given by

8.1 $$\delta' = b(k) + b(k+1) + \cdots + b(n),$$

where, as usual, $b(r)$ is the probability of r successes in n trials with success probability p_0. For alternatives $p < p_0$, the descriptive level is given by

8.2 $$\delta' = b(k) + b(k-1) + \cdot \cdot \cdot + b(0).$$

The descriptive level δ' characterizes how extreme our test result is in probability terms. It equals the significance level α' associated with a right- or left-tailed critical region whose cutoff point is k, where k is the actually observed number of successes:

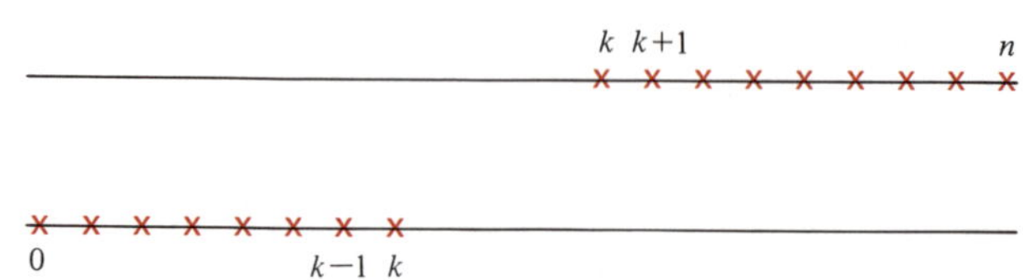

For a test of the hypothesis $p = p_0$ against the two-sided alternative $p \neq p_0$, the descriptive level δ'' is found by doubling the descriptive level of the appropriate one-sided test. Thus $\delta'' = 2\delta'$, where δ' is given by (8.1) if $k > np_0$ and by (8.2) if $k < np_0$.

EXAMPLE 8.3 For the extrasensory perception example in Section 51, we have $k = 8$, $n = 13$, and $p_0 = \frac{1}{2}$. Since we are interested in the one-sided alternative $p > \frac{1}{2}$, the descriptive level associated with the observed test result equals $b(8) + b(9) + \cdot \cdot \cdot + b(13) = .157 + .087 + \cdot \cdot \cdot + .000 = .291$. This is, of course, the significance level α' associated with the critical region $k \geq 8$ as given in Table 7.1. If Mrs. X has no extrasensory perception, there are roughly 3 chances in 10 that we would observe a result that is as extreme as the result that we actually did observe.

EXAMPLE 8.4 Assume that we observe three heads and seven tails in the coin-tossing experiment discussed in Section 58. Since the alternative is two-sided, $p \neq \frac{1}{2}$, we have $\delta'' = 2\delta'$, where $\delta' = b(3) + b(2) + b(1) + b(0) = .117 + .044 + .010 + .001 = .172$ or $\delta'' = .344$. In ten fair tosses of a coin, there is roughly one chance in three of observing a result that is at least as extreme as the (3,7)-split among heads and tails which we observed.

EXAMPLE 8.5 In a random sample of 144 students, 105 indicated that they would support a campus-wide demonstration. Is there evidence that fewer than four out of every five students are in favor of the demonstration? Let us compute the descriptive level corresponding to a test of the hypothesis $p = .8$ against the alternative $p < .8$. We have $\delta' = b(105) + b(104) + \cdot \cdot \cdot + b(0)$. Since the number of trials is large, we use the normal approximation to compute this sum of probabilities. We find that $\mu = np = 144 \times .8 = 115.2$ and $\sigma = \sqrt{npq} = \sqrt{144 \times .8 \times .2} = 4.8$, so that $z = (105 - 115.2)/4.8 = -2.12$. The lower tail probability associated with this value of z equals .017. Since $\delta' = .017$, it seems doubtful that at least 80 percent of the student body is in favor of the demonstration.

Example 8.5 illustrates the ease of finding the descriptive level with the help of the normal approximation. Against the alternative $p < p_0$, the descriptive level δ' is equal to the area under the standard normal

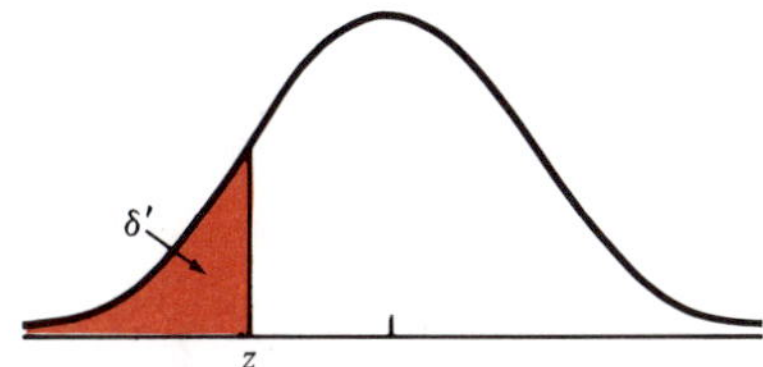

curve to the left of the z-value associated with the observed number of successes k, $z = (k - np_0)/\sqrt{np_0(1 - p_0)}$. Against the alternative $p > p_0$, the descriptive level δ' is equal to the area to the right of the appropriate z-value. We will see in later chapters that a similar approach works in many other cases.

65 Let us consider the following rule: Reject the hypothesis under consideration if the descriptive level associated with an experiment is smaller than or equal to α; do not reject the hypothesis under consideration if the descriptive level is greater than α. We will show that this rule provides a test with significance level α. Thus computation of the descriptive level allows us to test a given hypothesis without actually setting up a critical region.

For definiteness, we discuss the one-sided case where we are interested in testing the hypothesis $p = p_0$ against the alternative $p > p_0$. The other cases can be handled in similar fashion. According to our discussion in Section 62, the appropriate critical region with significance level α contains the values $(k_\alpha, k_\alpha + 1, \ldots, n)$, where

$$b(k_\alpha) + b(k_\alpha + 1) + \cdots + b(n) = \alpha.$$

Suppose that the number of successes actually observed is k. The hypothesis being tested is rejected if $k \geq k_\alpha$. According to (8.1), the descriptive level δ' associated with the observed value k equals

$$\delta' = b(k) + b(k + 1) + \cdots + b(n).$$

It follows that we can have $k \geq k_\alpha$ indicating that k is in the critical region if and only if $\delta' \leq \alpha$. This establishes the validity of the earlier statement. It was implied in our discussion of hypothesis testing that the actual determination of the significance level to be used in a testing situation is not really a statistical problem. Two different persons facing the same testing situation may well have different ideas about which significance level to use. In research reports we often read that on the basis of experimental evidence the hypothesis in question is rejected at, say, significance level .05. The author of such a report imposes his personal choice of a significance level on the reader. If instead the report states that the descriptive level is .033, the reader can decide for himself whether the experimental evidence warrants rejection or acceptance of the hypothesis being tested. Thus a reader who prefers a .02 rather than a .05 significance level would not want to reject the hypothesis on the basis of the given experimental evidence. Stating the descriptive level associated with a test result is usually much more informative than a mere statement to the effect that a given hypothesis should or should not be rejected at a given significance level.

Tests of Hypotheses and Confidence Intervals

66 The descriptive level may be looked upon as a measure of the implausibility of the hypothesis being tested in view of the experimental evidence: the smaller the descriptive level, the more implausible the hypothesis under consideration. But the descriptive level does not furnish information about possible alternatives. Such information is supplied by an appropriate confidence interval.

In Section 23 we found a confidence interval for the number of taxis by carrying out a series of tests of hypotheses. The confidence interval consisted of parameter values such that a test did not lead to rejection of the corresponding hypothesis. More formally, let θ be a parameter occurring in a statistical analysis. A confidence interval for θ contains all values θ_0 such that a test of the hypothesis $\theta = \theta_0$ does not lead to rejection of the hypothesis being tested. The confidence interval consists of "acceptable" θ-values or, reversing the argument, a value $\theta = \theta_0$ is said to be *acceptable* if it lies in the confidence interval for θ. We will often make use of this mutual relationship between tests of hypotheses on the one hand and confidence intervals on the other hand as we study new statistical procedures. Now we look at two examples involving the parameter p of a binomial distribution.

EXAMPLE 8.6 We return to the data for the experiment in Section 19 in which students were asked to mentally draw a sequence of two pingpong balls each labeled 1, 2, or 3. When we first saw the data in Table 3.21, even though we were relying on intuition, it seemed fairly evident that the number of students who selected the same digit twice was considerably lower than could be expected in a similar experiment with real pingpong balls. Now we have the tools for a more thorough investigation. For real pingpong balls the probability of getting the same digit twice in a row is $\frac{1}{3}$. Suppose that we let p be the probability that a person participating in our experiment puts down the same digit twice in succession. We can then test the hypothesis that $p = \frac{1}{3}$, the same as for draws of real pingpong balls. Rather than find a critical region and simply reject or accept the hypothesis $p = \frac{1}{3}$, it is more informative to find a confidence interval for p. To be really confident of our conclusions, let us make it a 99 percent confidence interval. The only bit of information we need is the number of successes. This we get from Table 3.21. There are 62 double-ones, 54 double-twos, and 58 double-threes. In all, then, the quantity k equals $62 + 54 + 58 = 174$. We will use confidence interval (6.9),

$$\frac{k}{n} - z\sqrt{\frac{(k/n)[1-(k/n)]}{n}} \leq p \leq \frac{k}{n} + z\sqrt{\frac{(k/n)[1-(k/n)]}{n}},$$

which for $k = 174$, $n = 900$, $z = 2.576$ becomes

$$.16 \leq p \leq .23.$$

We note that the value $p = \frac{1}{3}$ is not in the interval. Thus we may say that on the basis of available information, the value $p = \frac{1}{3}$ does not deserve our confidence. This, of course, is just another way of saying that we should reject the hypothesis $p = \frac{1}{3}$. What about the significance level of the corresponding test? Since we are 99 percent confident in our confidence interval statement, there is only 1 chance in 100 that we do not include the correct value p in the interval. In particular, if $p = \frac{1}{3}$ is the correct value, then there is only 1 chance in 100 that the value $\frac{1}{3}$ is not included in the confidence interval. The significance level α is 1 in 100, or .01.

The specific relationship between the confidence coefficient of a confidence interval and the significance level of the corresponding test that we have just found in Example 8.6 can be generalized. If γ is the confidence coefficient and α is the significance level, then we have

$$\alpha = 1 - \gamma.$$

EXAMPLE 8.7 In Section 58 we discussed tests of the hypotheses $p = \frac{1}{2}$ based on the results of $n = 10$ trials. The critical region with significance level .05 or less turned out to be $k = (0, 1, 9, 10)$. Thus at the .05 significance level, the hypothesis $p = \frac{1}{2}$ is rejected if in 10 trials we observe 0, 1, 9, or 10 successes. In all other cases the hypothesis $p = \frac{1}{2}$ cannot be rejected at the chosen significance level. The same (and more) information is revealed by Table 6.11 which, for $k = 0, 1, \ldots, 10$, lists confidence intervals for p with confidence coefficient at least .95 corresponding to k successes in 10 trials. The intervals for $k = 0, 1, 9,$

10 successes do not contain the value $p = \frac{1}{2}$, while the remaining intervals do contain the value $p = \frac{1}{2}$.

A confidence interval is usually much more informative than a test of a hypothesis. Assume, for example, that we have observed $k = 6$ successes in 10 trials. According to Example 8.7, the hypothesis $p = \frac{1}{2}$ is accepted by both the test based on the critical region $k = (0, 1, 9, 10)$ and the appropriate confidence interval, $.26 \leq p \leq .88$. However, the test provides information about the parameter value $p = \frac{1}{2}$ only. The confidence interval, on the other hand, shows that parameter values like $\frac{1}{3}$ and $\frac{3}{4}$ are also acceptable when we observe 6 successes in 10 trials. Indeed, the length of the confidence interval reveals that the result of 10 trials sheds very little light about the true value of p.

PROBLEMS

1 The following *triangle taste test* procedure is sometimes used to locate expert tasters. In the case of wine tasting, two glasses are filled with one kind of wine, a third glass is filled with a different kind of wine. The test subject is then asked to identify the single glass. If the two kinds of wine are very similar, an inexpert wine taster will simply guess and has then probability $\frac{1}{3}$ of locating the correct glass. An expert wine taster should be able to do much better.

Suppose that a test subject performs ten triangle tests and makes seven correct and three incorrect identifications. Would you say that the test subject is an expert wine taster?

a. Set up an appropriate hypothesis.
b. What is the alternative?
c. Find the descriptive level.

2 Find the descriptive level for Problem 15 of Chapter 7.

3 Find the descriptive level for Problem 24 of Chapter 7.

4 Find the descriptive level for Problem 26 of Chapter 7.

5 Find the descriptive level for Problem 27 of Chapter 7.

6 Find the descriptive level for Problem 28a of Chapter 7.

7 You have found the confidence interval $.68 \leq p \leq .73$ having confidence coefficient .95. Consider the following two statements.

(i) The hypothesis $p = \frac{2}{3}$ is accepted at significance level .05.
(ii) The hypothesis $p = \frac{3}{4}$ is accepted at significance level .05.

Which, if either, of these two statements is correct?

8 Find a confidence interval for the probability that a person has brown eyes using the data for Problem 6 of Chapter 10. Compare the confidence interval with the test result for Problem 16 of Chapter 7.

9 Find a confidence interval for the probability that a person has dark hair using the data for Problem 6 of Chapter 10. Compare the confidence interval with the test results for Problem 17 of Chapter 7.

10 Compare the confidence intervals for Problems 12 through 14 in Chapter 6 with the corresponding test results for Problems 18 through 20 in Chapter 7. Which approach provides more information, tests of hypotheses or the confidence intervals?

11 Find a confidence interval for the probability of white winning a game of chess using the data for Problem 22 of Chapter 6,

a. using all 200 games;
b. ignoring drawn games.

9

CHI-SQUARE TESTS

67 We have seen how to analyze the results of experiments in which each trial has only two possible outcomes. Now we want to take up the case when a trial has more than two possible outcomes. Here are a few examples.

In a table of random digits, there are 10 possibilities for each position, namely, the digits 0, 1, . . . , 9. When rolling two dice, the sum of the number of points showing can range from 2 to 12, with a total of 11 possibilities. At many colleges a student gets one of 5 possible grades, *A*, *B*, *C*, *D*, or *F*, in a course he takes for credit. A geneticist is interested in the various types of offspring that result from the crossing of certain parental traits.

It may sometimes be appropriate to reduce the number of categories to two and then apply one of the statistical analyses discussed earlier. Thus in the case of a student's grades, we may simply indicate whether the student passes or fails. But often such a simplification ignores important details. A passing grade may be given for outstanding work as well as for work that is barely satisfactory. We need methods for analyzing data resulting from trials with more than two outcomes or categories.

In general, the true probabilities associated with the various categories are unknown, but often we have certain ideas or hypotheses about what values these probabilities might or should have. Thus consider the following example. The State Police of a certain state have used the same number of patrols each day of the week to control traffic on state highways. Alarmed over an increasing number of fatal accidents, they wonder whether some modification in the number of patrols might not reduce the overall number of such accidents. On checking the records of 175 past accidents involving at least one fatality and classifying each such accident according to the day of the week on which the accident occurred, they come up with the data in Table 9.1. Is there sufficient evidence to conclude that fatal accidents are not spread evenly over the week?

TABLE 9.1

Day of the Week	S	M	Tu	W	Th	F	Sa
Number of Accidents	36	20	17	22	21	26	33

Suppose that we compare the actually observed frequencies of fatal accidents in Table 9.1 with frequencies *expected* under the assumption that accidents are spread evenly over the days of the week. Reasonable agreement between observed and expected frequencies would imply that no changes in the present traffic patrol pattern are called for, at least not for reasons of the present investigation. If there is considerable disagreement between observed and expected frequencies, suitable changes in the traffic patrol pattern may bring about a reduction in the overall number of accidents.

If fatal accidents occur equally often on any one day of the week, we should expect to have an average of $175/7 = 25$ accidents each day. The observed number of accidents seems to agree reasonably well with the expected number of accidents on Mondays, Wednesdays, Thursdays, and Fridays. Tuesday figures are rather low, and there seem to be marked excesses on Saturdays and Sundays. What is the overall picture? Can the observed deviations be explained in terms of chance fluctuations that occur in all such data, or is there something wrong with the original assumption that fatal accidents are evenly distributed over the days of the week? We need an objective criterion for making a decision. Such a criterion is developed in the next section.

The Chi-square Statistic

68 We are considering the following problem. We want to test a hypothesis about m probabilities $p_1, \ldots, p_m$ associated with m categories. Our data consists of observed frequencies $o_1, \ldots, o_m$. As long as these observed frequencies are "reasonably" close to theoretical or expected frequencies, $e_1 = np_1, \ldots, e_m = np_m$, where $n = o_1 + \cdots + o_m$ is the total number of observations, there is no reason to doubt the correctness of the hypothetical probabilities $p_1, \ldots, p_m$.

Statistical theory suggests the following criterion, usually denoted by χ^2 (chi-square), to measure the closeness between observed and expected frequencies:

9.2 $$\chi^2 = \frac{(o_1 - e_1)^2}{e_1} + \cdots + \frac{(o_m - e_m)^2}{e_m},$$

which can also be computed as

9.3 $$\chi^2 = \frac{o_1^{\,2}}{e_1} + \cdot \cdot \cdot + \frac{o_m^{\,2}}{e_m} - n.$$

Small values of χ^2 (that is, values of χ^2 near zero) suggest acceptance, and large values of χ^2 suggest rejection, of the hypothesis being tested. The test whose critical region consists of sufficiently large values of χ^2 is known as the chi-square test. This is one of the best known tests in all of statistics.

To carry out the chi-square test, we have to know how large is sufficiently large to allow rejection of the hypothesis that is being tested at a given significance level α. This information is obtained from a table of the chi-square distribution. Table D is such a table.

Before we can describe this new table, it is necessary to introduce a new term. Associated with every chi-square test is a quantity called the *number of degrees of freedom*. For the present test, this quantity is simply $m - 1$, one less than the number of categories involved. However, for other tests like the tests of association that we are going to discuss in the next chapter, the number of degrees of freedom is more complicated. The abbreviation *df* is often used to denote the number of degrees of freedom. In Table D degrees of freedom are listed in the left-hand margin. To test our hypothesis at significance level α, we use the row in Table D for $m - 1$ degrees of freedom and find the tabulated value in the column corresponding to the *upper tail* probability α. If the computed value of χ^2 exceeds the tabulated value, we reject the hypothesis.

It is shown in Section 71 that, for $m = 2$, the chi-square test is equivalent to the z-test of the hypothesis $p = p_o$ against the two-sided alternative $p \neq p_o$ based on (7.11) in Section 61. In Section 61 it was mentioned that the z-test assumes that the number n of trials is sufficiently large. The same requirement is necessary for the more general χ^2-test. This raises the question of how large n should be in practice before we feel justified in using the χ^2-procedure. This is a rather complicated problem that cannot be answered satisfactorily in a few words. However, the following somewhat overly conservative rule can provide guidance. The χ^2-procedure can be used satisfactorily for most practical purposes provided that each theoretical frequency e is at least 5.

69 We return now to the traffic accident example. Table 9.4 provides all the information required for the computation of χ^2. We see that $\chi^2 = 12$. (Note that in the present example there is no real need to compute the last column in Table 9.4. Since $e_1 = \cdot \cdot \cdot = e_7 = e$, say, χ^2 simplifies to $[o_1 - e)^2 + \cdot \cdot \cdot + (o_7 - e)^2]/e = 300/25 = 12$.)

We have $7 - 1 = 6$ degrees of freedom. Our value of χ^2 falls between the tabulated values of 10.6 and 12.6 corresponding to upper tail probabilities of .10 and .05, respectively. Thus the descriptive

TABLE 9.4

Categories	p	o	e	o − e	(o − e)²	(o − e)²/e
S	1/7	36	25	11	121	4.84
M	1/7	20	25	− 5	25	1.00
Tu	1/7	17	25	− 8	64	2.56
W	1/7	22	25	− 3	9	0.36
Th	1/7	21	25	− 4	16	0.64
F	1/7	26	25	1	1	0.04
Sa	1/7	33	25	8	64	2.56
Totals	1	175	175	0	300	12.00 = χ^2

level associated with our data lies between .05 and .10. An increased number of traffic patrols on Saturdays and Sundays, with corresponding reductions on Mondays, Tuesdays, Wednesdays, and Thursdays, may possibly save some lives.

70* We can give an intuitive explanation for the number $m - 1$ of degrees of freedom for the present chi-square test. To find the value of χ^2, we have to compute m expected frequencies $e_1 = np_1, \ldots, e_m = np_m$. However, since $p_1 = \cdots + p_m = 1$, we have $e_1 + \cdots + e_m = np_1 + \cdots + np_m = n(p_1 + \cdots + p_m) = n$, so that $e_m = n - (e_1 + \cdots + e_{m-1})$. Theoretically, it is sufficient to compute the $m - 1$ expected frequencies $e_1, \ldots, e_{m-1}$ according to the formula $e = np$ and then obtain e_m by subtraction. The important point is that e_m cannot be arbitrary; it is completely determined by the number of observations n and the remaining $m - 1$ theoretical frequencies $e_1, \ldots, e_{m-1}$. We therefore say that we have $m - 1$ degrees of freedom.

71* If there are only two categories, the testing problem of the present chapter reduces to the one discussed in Chapter 7. It is then of some interest to compare the chi-square test for $m = 2$ with the test procedure developed in Chapter 7. As a first step, we establish the appropriate relationship between the notation used in Chapter 7 and that of the present chapter. Let us refer to category 1 as success and to category 2 as failure and remember the notation p for the probability of success, $q = 1 - p$ for the probability of failure, and k for observed number of successes. Then we have

$$o_1 = k, \qquad o_2 = n - k, \qquad e_1 = np, \qquad e_2 = nq = n(1 - p),$$

so that

$$\chi^2 = \frac{(o_1 - e_1)^2}{e_1} + \frac{(o_2 - e_2)^2}{e_2}$$

$$= \frac{(k - np)^2}{np} + \frac{[(n - k) - n(1 - p)]^2}{nq}$$

$$= \frac{(k - np)^2}{np} + \frac{(np - k)^2}{nq} = \frac{(k - np)^2}{n}\left(\frac{1}{p} + \frac{1}{q}\right)$$

$$= \frac{(k - np)^2}{n}\frac{p + q}{pq} = \frac{(k - np)^2}{npq} = z^2,$$

where $z = (k - np)/\sqrt{npq}$ is the test statistic (7.11) used in Section 61 to test a hypothesis about the success probability p, when the number of trials is large.

Let c correspond to an upper tail probability α for the chi-square distribution with 1 degree of freedom appropriate for the case $m = 2$. Then the chi-square test rejects the hypothesis being tested if $\chi^2 > c$. However, since $\chi^2 = z^2$, this is equivalent to rejection provided that either

$$z > +\sqrt{c} \qquad \text{or} \qquad z < -\sqrt{c},$$

which is the critical region for the test based on z against a *two*-sided alternative. For example, for $\alpha = .05$, we find that $c = 3.84$ from Table D, and therefore $\sqrt{c} = \sqrt{3.84} = 1.96$ is the two-sided critical value for the normally distributed test statistic z.

Two Examples

72 *Grading on a Curve* In a course in elementary statistics attended by 283 students, the instructor gave the following grades: 25 *A*'s, 77 *B*'s, 114 *C*'s, 48 *D*'s, and 19 *F*'s. We want to find out whether this distribution of grades is in agreement with the probabilities p_A, p_B, p_C, p_D, and p_F derived in Example 5.30 for grading on a curve. The computations for χ^2 proceed as follows:

Grades	p	o	e	$o - e$	$(o - e)^2/e$
A	.07	25	19.8	5.2	1.37
B	.24	77	67.9	9.1	1.22
C	.38	114	107.6	6.4	0.38
D	.24	48	67.9	−19.9	5.38
F	.07	19	19.8	− 0.8	0.03
					8.83

Thus $\chi^2 = 8.83$ with four degrees of freedom. According to Table D, this value of χ^2 is not significant at the .05 significance level, but it is significant at the .10 level. A comparison of the actual and theoretical grade frequencies reveals that our instructor has been rather lenient in assigning grades.

How Not to Cheat According to our discussion, only sufficiently large values of χ^2 lead us to reject the hypothesis we are testing. But sometimes sufficiently small values of χ^2 also contain a message. Suppose that you are asked to test a certain die for fairness. More specifically, suppose that you are asked to roll the die 6000 times and note how often it falls 1, 2, 3, 4, 5, or 6. Rather than actually perform such a dreary experiment, some students may be tempted to invent "experimental" results. The data in Table 9.5 certainly look reasonable for a fair die.

TABLE 9.5 **Results of 6000 Imaginary Rolls of a Die**

Result of Roll	1	2	3	4	5	6
Number of Observations	991	1005	1015	994	1007	988

We will use the alternative formula (9.3) to compute χ^2. Since $e_1 = \cdot \cdot \cdot = e_m$, the formula is particularly simple:

$$\begin{aligned}\chi^2 &= (o_1{}^2 + \cdot \cdot \cdot + o_m{}^2)/e - n \\ &= (991^2 + 1005^2 + 1015^2 + 994^2 + 1007^2 + 988^2)/1000 - 6000 \\ &= .560.\end{aligned}$$

The resulting value of χ^2 is .560. This chi-square value is so small that there is only 1 chance in 100 that it may have arisen in a real experiment with a fair die. As always, statistical evidence is not conclusive. Such data may come out of a real experiment, but they do look very suspicious. Some data are just too good to be true. It takes considerable experience to fake data convincingly.

PROBLEMS

1 Suggest an appropriate hypothesis for the experiment in Problem 1 of Chapter 2 and use the data you obtained to test the hypothesis. Repeat for Problems 2 through 4 of Chapter 2.

2 Consider the data in Table 3.21.

a. Test the hypothesis that a student writes down the digits 1, 2, or 3 as the first digit with probability $\frac{1}{3}$ each.
b. Test the hypothesis that a student writes down each of the nine possible pairs (1,1), (1,2), . . . , (3,3) with probability $\frac{1}{9}$.

3 Repeat Problem 2 using the data in Table 3.22. Compare with results of Problem 2 and explain any possible differences.

4 In a famous experiment the geneticist Gregor Mendel found that a sample from the second generation of seeds resulting from crossing yellow round peas and green wrinkled peas could be classified as follows:

Yellow and round	315	*Green and round*	108
Yellow and wrinkled	101	*Green and wrinkled*	32

According to the theory proposed by Mendel, the corresponding probabilities are $\frac{9}{16}$, $\frac{3}{16}$, $\frac{3}{16}$, and $\frac{1}{16}$, respectively. Does the experiment bear out these theoretical probabilities?

5 Among the first 1000 digits of the number π, we find the following distribution of digits:

0	1	2	3	4	5	6	7	8	9
90	120	103	98	90	93	95	102	102	107

As far as frequency of occurrence is concerned, can these digits be considered random?

6 With the aid of various yearbooks and almanacs, a reader made a list of 1000 *first* digits encountered in various tables:

1	2	3	4	5	6	7	8	9
275	166	112	85	72	58	46	44	42

Intuitively, we would expect that every nonzero digit has an equal chance of turning up in a table of numerical facts. Test a corresponding hypothesis. Can you think of reasons why this hypothesis is false?

7 Carry out a formal test of the hypothesis suggested by Problem 5 in Chapter 2.

8 Prices of shares on the stock market are recorded to 1/8th of a dollar. We might then expect to find stocks selling at prices ending in

$$0 \quad \tfrac{1}{8} \quad \tfrac{1}{4} \quad \tfrac{3}{8} \quad \tfrac{1}{2} \quad \tfrac{5}{8} \quad \tfrac{3}{4} \quad \tfrac{7}{8}$$

with about equal frequency. On a certain day, 120 stocks showed the following frequencies:

$$26 \quad 8 \quad 15 \quad 9 \quad 22 \quad 12 \quad 19 \quad 9$$

Set up a formal hypothesis and test it. What possible alternative to the hypothesis of equally distributed final eighth does the data suggest?

9 At a certain college the grade distribution for all students receiving letter grades is as follows:

Grade	*A*	*B*	*C*	*D*	*F*
Percent	28	38	25	6	3

During the first ten weeks of instruction, a student can elect a pass–fail option for one of his courses. Under the pass–fail option the instructor still submits a letter grade, but the registrar converts *A*, *B*, *C*, and *D* to pass and *F* to fail. A course taken under the pass–fail option does not count in the computation of a student's grade-point average. A study of 1000 pass–fail elections shows the following letter grade distribution:

A	*B*	*C*	*D*	*F*
30	220	440	230	80

Would you say that students elect the pass–fail option to protect their grade-point average? Why?

10 According to the 1960 census, the proportion of United States families with 0, 1, 2, . . . children is as follows:

Number of children	0	1	2	3	4	5	6 or more
Proportion	.14	.17	.26	.18	.11	.06	.08

Problem 5 in Chapter 1 reports on a survey involving 100 families. Are the sample data in agreement with the census data? If not, how do they seem to differ?

11 A person making a large number of long-distance calls has found from past experience that the proportion of calls completed on the first try is about one in four. He then computes the following additional probabilities:

$$P(\text{Call requires two tries for completion.}) = \frac{3}{4} \times \frac{1}{4} = \frac{3}{16} \doteq .19$$

$$P(\text{Call requires three tries for completion.}) = \frac{3}{4} \times \frac{3}{4} \times \frac{1}{4} = \frac{9}{64} \doteq .14$$

$$P(\text{Call requires more than three tries for completion.})$$

$$= 1 - \left(\frac{1}{4} + \frac{3}{16} + \frac{9}{64}\right) = \frac{27}{64} \doteq .42$$

He keeps a record of the number of tries required to complete the next 100 calls with the following results:

Number of tries required for completion	1	2	3	more than 3
Number of calls	23	33	29	15

a. Set up an hypothesis corresponding to the preceding probabilities and test it.
b. Can you think of a reason why the hypothesis is rejected?

12 120 pairs of twins are classified according to the sex of the two twins as follows:

Two boys	*Two girls*	*One boy, one girl*
38	34	48

Test the following two hypotheses:

a. Probabilities for the three classifications follow a binomial distribution with $n = 2$ and $p = \frac{1}{2}$.
b. All three classifications have the same probability. [The second hypothesis arises from the example following Formula (3.20).]

13* Show that (9.3) is equivalent to (9.2).

10

TESTS OF HOMOGENEITY AND INDEPENDENCE

73 Chi-square tests can be used to solve more complex problems than the type of problem discussed in Chapter 9. In the grading on a curve example of Section 72, we wanted to know whether the frequencies of A's, B's, C's, D's, and F's corresponded to specified areas under a normal curve. A different type of problem arises when two or more instructors teach parallel sections of the same course. We may then want to know if a student has the same chance of getting grades of A, B, etc., irrespective of the section the student attends.

Or again, in a preelection poll we may be interested in a comparison of male and female voters. Suppose that we have selected random samples of 200 male and 100 female students at a university and inquired about their political leanings. Table 10.1 is a tabulation of the results. Are voting intentions the same for male and female students? The hypothesis that voting intentions of male and female students do not differ can be tested by means of a *test of homogeneity.*

TABLE 10.1 **Observed Frequencies**

	Male (M)	*Female (F)*	*Totals*
Democrats (D)	69	21	90
Republicans (R)	52	23	75
Independents (I)	79	56	135
Totals	200	100	300

The six frequencies in Table 10.1 suggest a chi-square type test. How do we find expected frequencies that reflect the hypothesis of homogeneity? The hypothesis of homogeneity states that male and female students have identical voting preferences. But this hypothesis does not associate specific probabilities, p_D, p_R and p_I, with the three classifications. On the other hand, since 90 of the 300 students in our two samples designated themselves as Democrats, we can *estimate* p_D as $^{90}/_{300} = .30$, with $^{75}/_{300} = .25$ and $^{135}/_{300} = .45$ serving as estimates for p_R and p_I. Expected frequencies for a chi-square test are obtained by multiplying these estimated probabilities by the appropriate sample sizes. For example, the male sample contains 200 students, and the expected number of males with Republican leanings is $200 \times .25 = 50 = 200 \times {}^{75}/_{300}$. More generally, the following formula provides expected frequencies:

10.2 $$\text{Expected frequency} = \frac{\text{column total} \times \text{row total}}{\text{total number of observations}}$$

Table 10.3 gives all six expected frequencies.

TABLE 10.3 **Expected Frequencies**

	Male	*Female*	*Totals*
Democrats	60	30	90
Republicans	50	25	75
Independents	90	45	135
Totals	200	100	300

The computation of χ^2 is now straightforward. Table 10.4 provides the details. Thus χ^2 equals 8.32. In Section 74 we see that the appropriate number of degrees of freedom associated with the present investigation is 2. According to Table D, rejection of the hypothesis of homogeneity is indicated (except at significance level .01 or smaller.) Male and female students do not seem to be homogeneous in their political leanings.

TABLE 10.4 **Computation of χ^2**

Sex of Student	*Political Leaning*	*o*	*e*	*o − e*	$\frac{(o-e)^2}{e}$
M	*D*	69	60	9	1.35
M	*R*	52	50	2	.08
M	*I*	79	90	−11	1.34
F	*D*	21	30	− 9	2.70
F	*R*	23	25	− 2	.16
F	*I*	56	45	11	2.69
					$8.32 = \chi^2$

Table 10.4 reveals very clearly the factors which make χ^2 large. According to the entries in lines 4 and 6, the observed number of female Democrats is considerably smaller, and the observed number of female Independents is considerably larger, than can be expected under the hypothesis of homogeneity.

A Test of Homogeneity

74 We will now describe in general terms the experimental setup that leads to a test of homogeneity. The experimenter is interested in a comparison of c experimental conditions, and makes

$$n = n_1 + \cdot \cdot \cdot + n_c$$

observations, n_1, under condition 1, . . . , n_c under condition c. An observation results from classifying a test subject as belonging to one of r categories. In the student poll example, $c = 2$ corresponding to male and female students; also $n_1 = 200$ and $n_2 = 100$. The possible categories are Democrats, Republicans, and Independents, so that $r = 3$.

The hypothesis of homogeneity states that the different experimental conditions have no bearing on the probability that a test subject belongs in a particular category. We arrange observed frequencies in a rectangular array, often called a *contingency table*, with c columns corresponding to the c experimental conditions and r rows corresponding to the r categories. To test for homogeneity, we find expected frequencies using formula (10.2) and then compute χ^2 in the usual fashion. The number of degrees of freedom is $(r - 1)(c - 1)$.

EXAMPLE 10.5 Suppose that we are interested in finding out if the choice of a student's major field of interest at a certain college is related to class level. We ask samples of 100 freshmen, sophomores, juniors, and seniors at the college for their major field of concentration with the results given in Table 10.6. This is a typical case of testing for homogeneity. Is the proportion of natural science (social science, humanities) majors the same for all four classes or are there differences? We use a chi-square test to find out.

TABLE 10.6

Field of Concentration	Class: Freshman	Sophomore	Junior	Senior	Totals
Natural Sciences	45	36	26	28	135
Social Sciences	25	33	41	38	137
Humanities	30	31	33	34	128
Totals	100	100	100	100	400

The expected number of freshmen who are natural science majors is $(100 \times 135)/400 = 33.75$. (While observed frequencies are of necessity integral numbers, it is bad practice to round off expected frequencies to the nearest integer.) Since all four samples are of the same size, the expected frequencies of sophomores, juniors, and seniors who are natural science majors are also 33.75. Similarly, we find the expected frequency of social science majors in any one class to be $(100 \times 137)/400 = 34.25$ and the expected number of humanity majors, $(100 \times 128)/400 = 32$. Thus

$$\chi^2 = \frac{(45 - 33.75)^2 + (36 - 33.75)^2 + (26 - 33.75)^2 + (28 - 33.75)^2}{33.75}$$

$$+ \frac{(25 - 34.25)^2 + (33 - 34.25)^2 + (41 - 34.25)^2 + (38 - 34.25)^2}{34.25}$$

$$+ \frac{(30 - 32)^2 + (31 - 32)^2 + (33 - 32) + (34 - 32)^2}{32}$$

$$= 6.66 + 4.28 + 0.31 = 11.25.$$

With $(r - 1)(c - 1) = (3 - 1)(4 - 1) = 6$ degrees of freedom, the result is not significant at the .05 level. If we adhere strictly to a 5 percent significance level, we have no reason to reject the hypothesis of homogeneity. However, if we adopt a purely exploratory point of view, the data certainly put us on guard that the hypothesis may not be strictly correct. A comparison of observed and expected frequencies suggests that the proportion of natural science majors may be decreasing, and the proportion of social science majors may be increasing, as students progress from freshmen to seniors. The proportion of humanities majors seems to be rather stable.

75 2 × 2 *Tables*

One of the most frequent applications of homogeneity testing occurs in the comparison of two binomial probabilities. Suppose that an investigator is interested in comparing the effectiveness of two medical treatments. Before he recommends one treatment in preference to the other, he will want to test the hypothesis that

$$p_1 = p_2$$

against the alternative that

$$p_1 \neq p_2,$$

where p_1 and p_2 are the recovery rates associated with the two treatments.

A test of homogeneity with $r = c = 2$ can be used to analyze appropriate data. The columns of the contingency table refer to the treatments; the rows refer to recovery or nonrecovery of patients when one or the other treatment has been applied. The resulting chi-square test has 1 degree of freedom. For 2×2 tables, χ^2 can be computed using a special formula. Suppose the contingency table is written as

			Totals
	a	b	a + b
	c	d	c + d
Totals	a + c	b + d	n = a + b + c + d

where a, b, c, d are the observed frequencies for the four cells. The following expression is equivalent to the usual formula for χ^2,

10.7 $$\chi^2 = \frac{n(ad - bc)^2}{(a + b)(c + d)(a + c)(b + d)}.$$

(Note that the four factors in the denominator are the two row and the two column totals.)

EXAMPLE 10.8 Of 110 patients who complained of headaches, 50 were given drug 1, and 60 were given drug 2. If the number of patients experiencing relief is 39 in the first group and 41 in the second group, can we conclude that there is a difference in the effectiveness of the two drugs? We construct the contingency table:

	Drug		
	1	2	Totals
Relief	39	41	80
No relief	11	19	30
Totals	50	60	110

Then we have

$$\chi^2 = \frac{110(39 \times 19 - 41 \times 11)^2}{50 \times 60 \times 80 \times 30} = 1.28.$$

This result does not indicate a significant difference in the effectiveness of the two drugs.

Formula (10.7) is useful for routine computations. It has the disadvantage that it does not require the computation of expected frequencies, which are most important for the interpretation of test results. If the value of χ^2 in Example 10.8 had turned out to be significant, we would still have the problem of deciding which treatment is more effective. This problem can be decided only by comparing observed frequencies with expected frequencies.

76 One-sided Alternatives

The chi-square tests that we have discussed are omnibus tests—that is, they test the stated hypothesis against the completely general alternative that the hypothesis is untrue. In Section 75 the hypothesis $p_1 = p_2$

is tested against the *two-sided* alternative $p_1 \neq p_2$. If either the number of rows or the number of columns in a contingency table is greater than 2, such general alternatives are usually appropriate. In the 2×2 case, however, specific alternatives are often more useful.

EXAMPLE 10.9 Samples of college students were asked the following question in 1970 and in 1973: Do you think that use of marijuana should be legalized? Answers are tabulated in the following contingency table:

	1970	*1973*	*Totals*
Yes	53	141	194
No	47	59	106
Totals	100	200	300

Has student opinion concerning legalization of marijuana changed between 1970 and 1973? Let p_{70} denote the proportion of students who favored legalization of marijuana in 1970 and p_{73}, the corresponding proportion for 1973. We set up the null hypothesis $p_{73} = p_{70}$, which states that no change has taken place. Even without looking at the data—and we should never look at the data when deciding on an appropriate alternative—we may feel that, if there has been any change, it would be an upward change. The alternative is then one-sided, $p_{73} > p_{70}$, and the chi-square test of Section 75 is inappropriate. The following modification converts the regular chi-square test for 2×2 tables into a test against a one-sided alternative. We compute expected frequencies in the usual way. If the observed frequencies deviate from the expected frequencies in the manner predicted under the alternative, we compute χ^2. The descriptive level of the one-sided test is given by the area under the standard normal curve to the right of $\sqrt{\chi^2}$.

EXAMPLE 10.9 *cont'd.* For the data of Example 10.9, we find the following expected frequencies:

	1970	*1973*
Yes	64.7	129.3
No	35.3	70.7

For 1970 the observed number of students who favor legalization of marijuana is 53. This is smaller than the number, 64.7, expected under the null hypothesis. For 1973, the corresponding observed frequency is 141. This is greater than 129.3 expected under the null hypothesis. The deviations are in agreement with the alternative $p_{73} > p_{70}$. Using (10.7), we find that $\sqrt{\chi^2} = \sqrt{8.94} = 2.99$, which corresponds to a descriptive level of .001 and is a highly significant result.

A Test of Independence

77 An $r \times c$ contingency table may arise from an investigation which differs in its experimental setup from the one discussed in Section 74. We consider the following example. An insurance company writing automobile insurance is interested in a possible relationship between the smoking habits of the principal operator of an automobile and the frequency of accident claims for property damage. The company classifies each policyholder in one of three categories: low, medium, or high accident frequency. Some of the company's automobile policyholders also carry life insurance policies with an affiliate company that classifies its policyholders (among others) as smokers and nonsmokers. A random sample of 600 policyholders who carry both automobile and life insurance show the breakdown in Table 10.10.

TABLE 10.10 Smoking and Automobile Accidents

Accident Frequency	*Smokers (S)*	*Nonsmokers (N)*	*Totals*
Low (L)	35	170	205
Medium (M)	79	193	272
High (H)	57	66	123
Totals	171	429	600

The computation of χ^2 proceeds as follows:

Row	*Column Category*	*o*	*e*	$(o - e)^2/e$
L	*S*	35	58.4	9.4
L	*N*	170	146.6	3.7
M	*S*	79	77.5	0.0
M	*N*	193	194.5	0.0
H	*S*	57	35.1	13.7
H	*N*	66	87.9	5.5
				$\chi^2 = 32.3$

For 2 degrees of freedom, the observed value of chi-square is highly significant suggesting that smokers and nonsmokers have different accident frequencies. In the low accident group, only one policyholder in six is a smoker; in the high accident group, the ratio is nearly one in two. The insurance company may well feel justified in offering lower property damage rates to nonsmokers than to smokers.

While we have analyzed our data exactly as for a test of homogeneity, in reality we are dealing with a different experimental setup. For a test of homogeneity we would require separate samples of smokers and nonsmokers. In the present investigation it was more convenient to select a single sample of policyholders with policies in both companies and to classify each member of the sample with respect to two characteristics, smoking habit and accident frequency.

The chi-square test now becomes a test of the hypothesis that the two characteristics, smoking habit and accident frequency, are independent.

78* The distinction between χ^2 as a test of homogeneity and a test of independence becomes clearer by stating the problem in more general terms. A potential test subject can be classified according to two characteristics A and B. Characteristic A has r categories $A_1, \ldots, A_r$; characteristic B has c categories $B_1, \ldots, B_c$. In a test of independence we obtain n test subjects and classify each according to the A-category and the B-category to which he or she belongs. If we assign characteristic A to rows and characteristic B to columns, we find that observed frequencies form a contingency table with r rows and c columns. We say that characteristics A and B are independent if for all A-categories, A_i, and all B-categories, B_j, we have

10.11

$$P(A_i \text{ and } B_j) = P(A_i)P(B_j).$$

When we use the chi-square test as a test of independence, we are testing the validity of (10.11).

We have a different type of problem when we use characteristic B to divide potential test subjects into c groups corresponding to the category B_j to which the test subject belongs. In a test of homogeneity, instead of selecting one sample of size n from among all test subjects as in a test of independence, we select c samples, one from each of the c groups that correspond to categories $B_1, \ldots, B_c$. The hypothesis of homogeneity is then stated as follows. If A_i is any A-category, we have

$$P_{B_1}(A_i) = \cdots = P_{B_c}(A_i),$$

where the subscripts $B_1, \ldots, B_c$ indicate that we restrict ourselves to test subjects that have the indicated B-classification.

If in the college poll example of Section 73 we set $B_1 = M =$ male, $B_2 = F =$ female; $A_1 = D =$ Democrat, $A_2 = R =$ Republican, $A_3 = I =$ Independent, the formal hypothesis of homogeneity becomes

$$\begin{aligned} &P_M(D) = P_F(D) \\ \text{and}\quad &P_M(R) = P_F(R) \\ \text{and}\quad &P_M(I) = P_F(I). \end{aligned}$$

The chi-square test in Section 73 raises doubts about the validity of the first and third of these equalities.

79* We conclude this chapter by indicating why, for an $r \times c$ contingency table, the appropriate number of degrees of freedom is $(r - 1)(c - 1)$. In Section 68 we had $m - 1$ degrees of freedom for the simple chi-square test in view of the fact that the restriction $e_1 + \cdots + e_m = o_1 + \cdots + o_m = n$ allowed us to compute one of the expected frequencies, say e_m, by subtraction. In the case of an $r \times c$ contingency table, expected frequencies in every row and in every column add up to the same totals as the observed frequencies. As a consequence, expected frequencies in the last row and in the last column can theoretically be computed by subtraction. This leaves only $(r - 1)(c - 1)$ expected frequencies to be computed according to (10.2). Thus the present chi-square test has $(r - 1)(c - 1)$ degrees of freedom.

PROBLEMS

In problems where the hypothesis being tested is rejected, the student should indicate what alternative is suggested by the data.

1 At a university, students from two different colleges take the same calculus examination. Here are the grades:

	Grades				
	A	*B*	*C*	*D*	*F*
College 1	6	13	43	16	22
College 2	18	26	41	6	9

Set up an appropriate hypothesis and test it.

2 Fifty smokers and 50 nonsmokers are asked whether they believe that heavy smoking may lead to lung cancer and other serious diseases. Their answers are tabulated as follows:

	Do believe	*Do not believe*
Smokers	11	39
Nonsmokers	28	22

Set up an appropriate hypothesis and test it.

3 In a preelection poll studying the influence of age on voter preference for two presidential candidates, the following results were obtained:

	Prefer candidate A	*Prefer candidate B*	*Undecided*
20–29	67	117	16
30–49	109	74	17
Over 49	118	64	18

Set up an appropriate hypothesis and test it.

4 The data in Table 10.6 can be arranged as follows:

	Freshman	*Sophomore*	*Junior*	*Senior*
Other than humanities	70	69	67	66
Humanities	30	31	33	34

Test the hypothesis that the choice of a humanities major is unrelated to the class of the student.

5 A final examination was taken by 105 male students and 40 female students. The instructor noted whether a student was sitting in the front rows or the back rows:

	Front rows	*Back rows*
Male students	35	70
Female students	20	20

Set up an appropriate hypothesis and test it.

6 A group of 100 persons has been classified according to eye and hair color:

		Eye color:		
		Brown	*Blue*	*Grey*
Hair color:	*Light*	13	18	9
	Dark	37	12	11

Would you say that eye and hair color are related?

7 In a preelection poll 500 each of three groups of students from different fields of study were asked their political preference:

	Field of study:		
	Humanities and social sciences	*Physical sciences and engineering*	*Professional schools*
Democrat	305	230	215
Republican	195	270	285

Can we conclude that political preference is related to field of study?

8 The following table gives the number of children with dental cavities in surveys conducted before and after fluoridation of a city's water supply:

	Before fluoridation	*After fluoridation*
With cavities	143	62
Without cavities	57	88

Set up an appropriate hypothesis and test it.

9 In an experiment to investigate the effectiveness of cloud seeding, the following data were obtained:

	Seeded days	*Unseeded days*
Precipitation	42	34
No precipitation	61	63

Can we conclude that cloud seeding has altered the frequency of days with precipitation?

10 In a survey college seniors were asked whether they planned to go on for their doctorates. Of 1000 white seniors, 140 answered in the affirmative. Of 100 black seniors, 18 answered in the affirmative. Is there a difference between the aspirations of white and black seniors to work for a doctorate?

11 Three instructors taught parallel sections of the same statistics course:

		Grades:				
		A	*B*	*C*	*D*	*F*
	K	25	25	66	45	20
Instructor:	*N*	13	30	53	36	16
	P	18	30	43	32	9

Set up an appropriate hypothesis and test it.

12 The following question was asked of samples of college students in 1969 and 1973: Are casual premarital sexual relations morally wrong? The answers are recorded below:

	1969	*1973*
Yes	68	66
No	132	234

Have the attitudes of students changed between 1969 and 1973?

13 One hundred executives were interviewed about their reading habits of two financial publications *A* and *B*:

	Read A	*Do not read A*
Read B	65	5
Do not read B	17	13

Is readership of one publication related to readership of the other publication?

14 120 pairs of twins are classified according to the sex of the first and the second born twin:

		First born:	
		Male	*Female*
Second born:	*Male*	38	26
	Female	22	34

Test for independence of the sex of twins.

15 An executive personnel agency classified 1000 of its clients according to weight and income. Of 880 executives with incomes between 10000 and 20000 dollars, 352 were overweight. Of 120 executives with incomes above 20000 dollars, 11 were overweight. What conclusions do these data suggest?

16 To predict the outcome of a local bond issue referendum, 300 registered voters were contacted. In addition to the bond question, voters were asked whether they were registered as Democrats or Republicans or whether they were without party affiliation:

	Democrats	*Republicans*	*Unaffiliated*
Favor bond issue	56	32	72
Do not favor bond issue	30	39	28
No opinion	10	9	24

Set up an appropriate hypothesis and test it.

17 The following question was asked in three surveys of college student opinion: Should a woman be allowed to have an abortion—regardless of circumstances, under certain circumstances, under no circumstances? For the years 1970, 1971, and 1973, the answers were as follows:

	1970	*1971*	*1973*
Regardless of circumstances	472	645	690
Only in certain circumstances	458	314	275
Under no circumstances	70	41	35

Has there been a change of opinion over the years of the survey?

18 To investigate the influence of prejudicial news coverage of pretrial evidence on jury verdicts, the following experiment was conducted. 120 persons were selected from regular jury pools to attend the reenactment of a murder trial. Before the start of the trial, half of the "jurors" read prejudicial news stories while the other half read objective reports. After the trial, 47 persons in the first group and 33 persons in the second group voted for conviction. What conclusions do you draw from these figures?

11

THE STATISTICAL ANALYSIS OF QUANTITATIVE DATA

80 The methods of Chapter 6–10 are appropriate for the analysis of data of a categorical or qualitative nature. Quantitative measurements present different kinds of problems, some of which are discussed in subsequent chapters. But first some groundwork is needed.

Problems involving categorical data revolve around a limited number of probabilities associated with the individual categories. Phenomena like the number of miles an automobile runs on a gallon of gasoline, the weight of an individual, or the number of hours of television viewing provided by a television tube are not adequately described by a limited number of probabilities.

If you ask an individual how much he weighs, you are likely to get an answer like 143, or 160, or 275, pounds. We could then try to define weight categories like . . . 142, 143, 144 . . . and assign appropriate probabilities to these individual categories. One immediate difficulty arises from the large number of categories. More importantly, a weight like 143 is not exact but rounded. If a person says his weight is 143, he implies that his true weight is somewhere between $142\frac{1}{2}$ and $143\frac{1}{2}$ pounds. People have a tendency to round off their weight to the nearest pound or possibly even to the nearest 5 or 10 pounds. We then have to resolve the question of whether we should assign probabilities to five-pound intervals, one-pound intervals, or possibly half-pound intervals. Fortunately, there is no need to worry about this problem. Distribution functions, like the normal distribution, assign probabilities to all possible intervals, small or large.

When discussing binomial experiments in Section 27, we represented probabilities as areas under a histogram and then approximated the histogram by a normal distribution. We also pointed out that the significance of the normal distribution as a tool for computing probabilities extends far beyond problems involving binomial trials. For example, we used the normal distribution to compute probabilities relating to examination grades. Experience shows that normal distributions provide adequate descriptions for many kinds of measurements like weights of persons, gasoline mileage, and test scores.

There was a time when it was believed that measurements that could not be described by a normal distribution were abnormal and therefore suspect. Today we realize that this is not the case. Certain phenomena can be adequately described using a normal distribution, others cannot. For example, yearly family income is described by a curve like the one shown in Figure 11.1. Most families have incomes that fall in a relatively limited range, but a small proportion of families have incomes spreading far beyond this range. A distribution like Figure 11.1 is used

FIGURE 11.1

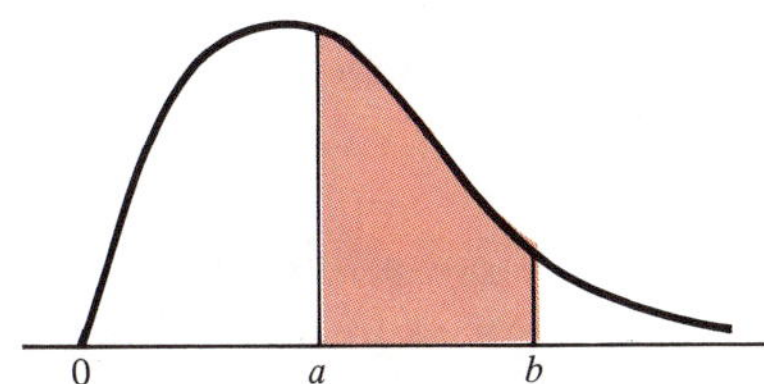

exactly like a normal distribution. The area under the curve between two ordinates a and b equals the proportion of families with yearly incomes between a and b dollars. Quite generally, we will assume that relevant probabilities are given by areas under appropriate distribution functions.

Nonparametric and Normal Theory Methods

81 Only rarely is the equation of the appropriate distribution function completely known. Often theory or past experience suggests a distribution function of a certain type, for example, the normal type, without specifying the exact values of parameters like μ and σ that enter into the equation. Much of present-day statistical methodology deals with estimation and hypothesis testing involving normal distributions. For experienced statisticians, such normal theory methods are immensely valuable. For beginners, they produce misleading results quite easily. We may compare normal theory methods to an expensive camera. In the hands of an experienced photographer, the camera produces results that are worth the purchase price. But beginners may get confused and discouraged. In most cases beginners would be much better off with a simple box-type camera.

We study normal theory methods in Chapters 17–20. But first we take up some "box-type" procedures. These procedures are simpler than normal theory methods. They afford the user more latitude in his assumptions. The word *robust* has been used to describe certain of their properties. Most commonly they are known as *nonparametric* methods even though sometimes they are concerned with parameters.

There is some danger of reading too much into our comparison of statistics and photography. Skilled photographers hardly ever use box cameras. Skilled statisticians regularly use nonparametric methods. Both normal theory and nonparametric methods have their rightful place in a statistician's tool kit. On the whole, inexperienced statisticians are less likely to go wrong when they use nonparametric methods than when they use normal theory methods.

The statement that nonparametric methods afford the user greater latitude in assumptions requires elaboration. When discussing the taxi problem in Chapter 1, we pointed out that statistical methods are not universal in their applications. Our estimate of the number of taxis could be expected to give useful information only if taxis were numbered consecutively starting with 1 (aside from the assumption of randomness of our sample, which is implicit for all methods discussed in this book). If our assumptions are inappropriate, our estimate may be greatly in error as in the example in Section 5.

The normal theory methods to be discussed in Chapters 17–20 are based on the assumption that observations can be described adequately by means of normal distributions. If this assumption is correct, normal theory methods are the best we can use. No other methods provide more accurate estimates of unknown parameters or smaller error probabilities in tests of hypotheses.

However, not all measurement data follow normal distributions. Clearly any asymmetry implies nonnormality. Even symmetry does not offer assurance of normality. The nonparametric methods discussed in the next five chapters are appropriate for all kinds of data. Certain theoretical difficulties may arise when nonparametric methods are applied to data that are not measured on a continuous scale (see Section 96). However, these theoretical difficulties need not concern us to any great extent.

Since the assumption of normality implies continuous measurements, nonparametric procedures are valid whenever normal procedures are. The reverse is not true. There are situations when the use of nonparametric methods is appropriate, while the use of normal theory methods may produce rather misleading results.

A Look Ahead

82 A population is completely described by its distribution function. But complete knowledge of the distribution is rarely necessary. Often a knowledge of one or two numerical constants that characterize important aspects of the distribution is all we need. In many investigations some kind of central value is of primary interest. Hence the question arises of what we mean by the center of a distribution.

For symmetric distributions the answer is obvious. We mean the center of symmetry. For asymmetric distributions a unique answer is not possible. What constitutes an appropriate measure of centrality depends on the purpose of an investigation. One of the simplest and most useful measures of centrality is the *population median* which divides the distribution in two equal parts, the lower and upper 50 percent. We denote the population median by the letter η as shown in Figure 11.2.

FIGURE 11.2

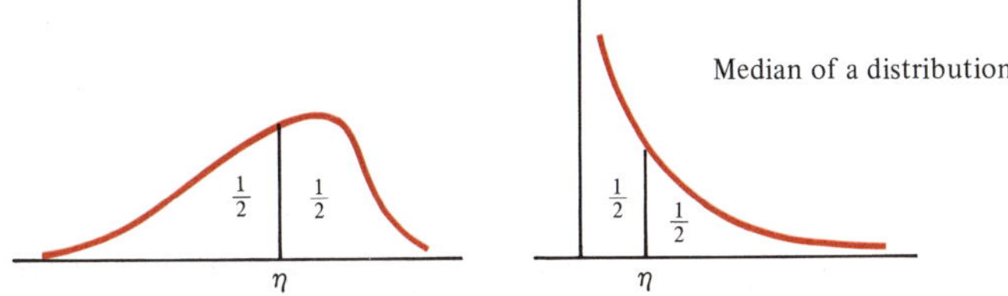

EXAMPLE 11.3 Government reports often quote the median income. The median income divides all incomes in two equal parts. There are as many families with income higher than the median as there are families with income lower than the median.

In Chapter 12 we find point and interval estimates for the median η of an unknown population and discuss hypothesis testing. In Chapters 13, 14, and 15 we extend considerations from a single population to two or more populations. In Chapter 16 we generalize the discussion of Chapter 10 on dependence or independence of categorical data to measurement data.

PROBLEMS

1 For each of the distributions in Figure 11.2, suggest a possible population.

2 Show that for symmetric distributions the population median coincides with the center of symmetry.

3 Find the population median for the following three distributions:

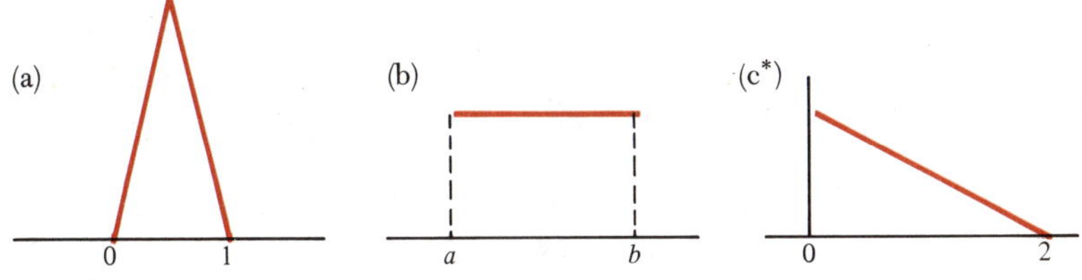

12

ONE-SAMPLE METHODS

83 Consider the following problem. We are interested in the median η of a certain population. The appropriate distribution function is unknown, but we observe a random sample of n observations which we denote by $x_1, \ldots, x_n$. We want to use these observations to estimate η, first by a point estimate and then by a confidence interval.

Specifically, we might think of the following situation. We should like to know something about the earnings of students at a large university. Because of the large number of students involved, it is impractical to ask each one about his earnings. By the method of Section 20, we select a random sample of, say, 100 students and find out what their earnings are. In this case, $n = 100$, x_1 denotes the earnings of the first student in our sample, x_2, those of the second student, and so on.

A rather obvious point estimate of the population median η is the sample median M, the median of the set of observations $x_1, \ldots, x_n$. In the student earnings example, we would arrange the various earnings from the lowest to the highest and then find the number halfway between the 50th and 51st observation in this rearranged series.

Confidence Intervals for Medians

84 Our point estimate of η has the usual disadvantages of point estimates in general. Information of its accuracy is frequently lacking. On the other hand, it is very easy to find a confidence interval for η, and the length of a confidence interval provides a built-in indicator of its accuracy. A short interval provides precise information; a long interval provides only vague information.

By far the simplest interval is obtained by using the smallest and the largest observations in the sample as lower and upper limits, that is, the interval

$$S_1 \leq \eta \leq L_1,$$

where S_1 is the smallest sample observation and L_1, the largest. What is the confidence coefficient γ associated with this interval? The answer to this question is quite simple. Our statement is correct unless our n observations are either all greater than the population median η or are all smaller than η. Since the median η divides the population in two equal parts, the probability that a single observation is greater than η is $\frac{1}{2}$. Since we assume that successive observations are independent, it follows from the multiplication theorem (3.17) that the probability that all n observations are greater than η is $(\frac{1}{2})^n$. This is also the probability that all observations are smaller than η. Thus

$$\gamma = 1 - 2\left(\frac{1}{2}\right)^n = 1 - \frac{1}{2^{n-1}}.$$

Table 12.1 illustrates this formula. We can be at least 99 percent confident that the population median lies between the smallest and the largest observations in a sample containing eight observations. If we are satisfied with a 97 percent confidence coefficient, we need only six observations.

Table 12.1 Confidence Coefficients of Confidence Intervals for Median

n	2	3	4	5	6	7	8
γ	.500	.750	.875	.938	.969	.984	.992

By taking samples containing more than eight observations, we can be even more confident in the correctness of our statements. However, there is rarely any practical need for having a confidence interval with confidence coefficient greater than, say, .99 or .995, and, of course, we must buy this high assurance of the correctness of our statement at some expense. As the number of observations increases, our method produces confidence intervals with higher and higher confidence coefficients. At the same time the length of the confidence interval tends to increase together with the number of observations. The smallest and the largest observations in a sample of size 8 are likely to be considerably farther apart than the smallest and largest observations in a sample of size 4. While a higher confidence coefficient means a more reliable statement, a longer interval means a less practically useful statement. We have noticed similar situations before in regard to confidence intervals for the parameter p of a binomial distribution and the inverse relationship between probabilities of type 1 and type 2 errors in testing statistical hypotheses. In determining a confidence interval for any parameter, we have to strike a balance between the confidence coefficient we should like to achieve and the expected length of the resulting confidence interval. What, then, can we do in this case?

The confidence interval bounded by the smallest and largest sample observations is as simple an interval as we can find. But almost as simple is an interval bounded by the second smallest and second largest observations in the sample. More generally, let

$$S_d = d\text{th smallest sample observation,}$$
$$L_d = d\text{th largest sample observation.}$$

We can then consider confidence intervals of the type

12.2 $$S_d \leq \eta \leq L_d.$$

The computation of the corresponding confidence coefficient γ is only slightly more complicated than in the case of $d = 1$, which we considered earlier. However, we will simply use Table E which, for various values of d, gives the confidence coefficient γ associated with the confidence interval (12.2). We will refer to intervals of type (12.2) as S-intervals.

NOTE 12.3 Many statisticians prefer routine confidence coefficients such as .90, .95, or .99. In general, the methods of this and the next two chapters do not permit us to find intervals corresponding exactly to one of these confidence coefficients. For this reason Table E (as well as Tables F and G) lists d-values such that the corresponding confidence coefficients bracket routine levels .99, .95, and .90. Statisticians can then select the d-value best suited to their purposes.

EXAMPLE 12.4 Nine randomly selected students obtained the following scores on a certain examination:

77, 88, 85, 74, 75, 62, 80, 70, 83.

Find a point estimate and confidence interval for the median score for this examination. The sample median is 77. This is our point estimate. According to Table E, a confidence interval with confidence coefficient .961 is bounded by the second smallest and the second largest of the observations. Thus we find the confidence interval

$$70 \leq \eta \leq 85.$$

As grades go, this is a rather wide interval, indicating that a sample of size 9 does not provide very precise information.

85 Table E provides d-values for sample sizes up to 50. For samples with more than 50 observations, the following normal approximation should be used:

$$d \doteq \frac{1}{2}(n + 1 - z\sqrt{n}),$$

where z is obtained from Table C according to the confidence coefficient γ (or, in case of hypothesis testing, according to the significance level α) that we want to achieve. For example, we have $n = 100$ observations and should like to find a confidence interval with confidence coefficient .99. Table C gives $z = 2.576$ corresponding to $\gamma = .99$ and

$$d \doteq \frac{1}{2}(100 + 1 - 2.576\sqrt{100})$$

$$= \frac{1}{2}(101 - 25.76) = 37.62.$$

We would then use $d = 38$.

86 *Symmetric Populations*

The confidence intervals for medians that we have just discussed are typical of nonparametric methods with respect to their conceptual and computational simplicity. More important, however, the results are valid without any assumptions about the population distribution. If special knowledge about the population distribution is available, more accurate methods are often possible. Thus it may be reasonable to assume that the population distribution is symmetric. The median η then coincides with the center of symmetry, and in finding an estimate for η, it is often advantageous to make use of symmetry.

For symmetric populations, the sample observations

12.5 $$x_1, x_2, \ldots, x_n$$

are symmetrically positioned around η. The same is also true of the $n(n - 1)/2$ averages of two observations

12.6 $$\frac{x_1 + x_2}{2}, \frac{x_1 + x_3}{2}, \ldots, \frac{x_{n-1} + x_n}{2}.$$

Since each of the observations in (12.5) can be written as $x_i = (x_i + x_i)/2$, we will refer to the combined $n + n(n - 1)/2 = n(n + 1)/2$ quantities (12.5) and (12.6) as *data averages*. For symmetric populations point estimates and confidence intervals for the center of symmetry η can be found in much the same manner as in Sections 83 and 84 for the median of an arbitrary population. As a point estimate of η, we use the median of all $n(n + 1)/2$ data averages. As the lower confidence limit, we use the dth smallest data average; as the upper limit, the dth largest data average, where this time Table F is used to relate confidence coefficients and d-values. We will refer to the resulting interval as the W-interval to distinguish it from the S-interval.

EXAMPLE 12.4 *cont'd.* Suppose that that past experience has told us that examination scores like the ones in Example 12.4 tend to be symmetrically distributed. We can then apply the new procedures to the data of Example 12.4. Let us first find a confidence interval. According to Table F, a confidence interval with confidence coefficient .961 is bounded by the 6th smallest and 6th largest of all data averages. Using either Table 12.7 or the graphical method of Section 88, we find that the 6th smallest average is 70, while the 6th largest is 84. Thus our confidence interval is $70 \leq \eta \leq 84$, slightly shorter than the earlier interval. But it should be remembered that we are making use of an additional assumption, namely, the symmetry of the distribution of grades. If the assumption of symmetry is unjustified, we have no right to use the second confidence interval.

The point estimate is given by the median of the $\frac{1}{2} \times 9 \times 10 = 45$ averages. Thus we are looking for the 23rd smallest (or largest) average, which is found to equal 77.5.

The $\frac{1}{2}n(n + 1)$ data averages are best exhibited in a table like Table 12.7, where the observations of Example 12.4 are entered along the diagonal, arranged from the smallest to the largest, and where, for example, the number 80 is the average of the corresponding two diagonal entries 75 and 85.

TABLE 12.7

								88
							85	86.5
						83	84	85.5
					80	81.5	82.5	84
				77	78.5	80	81	82.5
			75	76	77.5	79	80	81.5
		74	74.5	75.5	77	78.5	79.5	81
	70	72	72.5	73.5	75	76.5	77.5	79
62	66	68	68.5	69.5	71	72.5	73.5	75

87*

The following arithmetic scheme allows us to write down all $n(n + 1)/2$ data averages with a minimum of effort. To simplify the presentation, we use only the four smallest observations of Example 12.4: 74, 75, 62, 70.

The first step consists in arranging the observations according to size and finding *increments* equal to *half* the difference of two successive numbers:

Observations	62	70	74	75
Increments		4	2	0.5

Thus $2 = (74 - 70)/2$. We obtain our bottom row of averages by adding the increments successively to the smallest observation:

$$62 \qquad 62 + 4 = 66 \qquad 66 + 2 = 68 \qquad 68 + 0.5 = 68.5.$$

The row above the bottom row is obtained by adding the first increment —that is, 4—to all entries in the bottom row except to the element farthest to the left:

$$66 + 4 = 70 \qquad 68 + 4 = 72 \qquad 68.5 + 4 = 72.5.$$

Successive rows are found by continuing in the same fashion and using successive increments:

$$72 + 2 = 74 \qquad 72.5 + 2 = 74.5$$
$$74.5 + 0.5 = 75$$

Omitting the computational details, we then have all ten averages

			75
		74	74.5
	70	72	72.5
62	66	68	68.5

As a check on the correctness of our computations, the numbers in the diagonal from the lower left to the upper right should be the original observations arranged in increasing order.

88 The method of generating all data averages described in Section 87 is reasonably fast and simple as long as the number of observations involved is small. However, it has two disadvantages that become increasingly evident with increasing sample size. We have to compute all averages even though in the case of confidence intervals we are only interested in the more extreme averages. Also, the resulting table of averages is only partially ordered, namely, within each individual row and each individual column. When looking for the 6th smallest average, as in Example 12.4, we have to watch successive rows. The following graphic method possesses neither of these disadvantages.

On a sheet of graph paper we plot the data points along the 45° line starting at the lower left-hand corner. (A simple method for doing so is to plot the data points along the vertical axis on the left and project the resulting points horizontally onto the 45° line.) Below the 45° line we mark all intersections of parallels to the horizontal and vertical axes through the n data points. These intersections together with the data points on the 45° line represent the data averages. The intersections are ordered according to increasing (decreasing) averages by sliding a line perpendicular to the 45° line from below (above) and counting intersections *including* data points. A draftsperson's triangle made of clear plastic is very helpful in performing the counting operation.

The necessary steps for the graphic solution of Example 12.4 are indicated in Figure 12.8. Sliding the perpendicular to the 45° line from below, we find that the 6th intersection corresponds to the data point

FIGURE 12.8 Estimation of Center of Symmetry

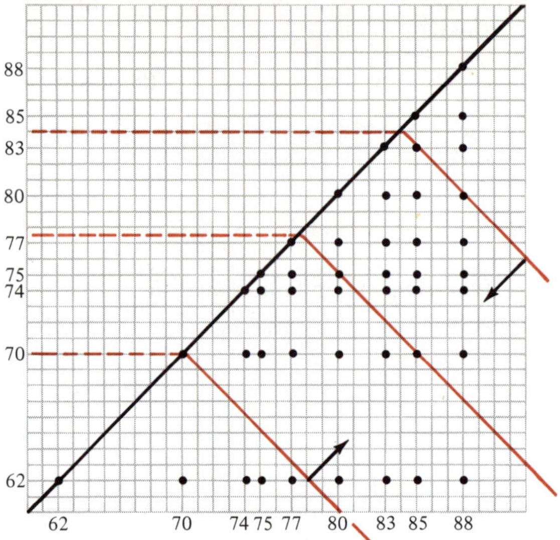

70. Thus the lower confidence limit is 70. Sliding the perpendicular from above, we find that the 6th intersection is formed by the horizontal through 80 and the vertical through 88. Thus the average equals $(80 + 88)/2 = 84$. This is the upper confidence limit. We obtain the same result if we use the intersection of the horizontal through 83 and the vertical through 85. The point estimate of η corresponds to the 23rd intersection from above or below. Again we have two choices: $(70 + 85)/2 = 77.5 = (75 + 80)/2$. The data average that corresponds to any particular point is found most easily by projecting the foot of the perpendicular through the point onto the vertical axis as shown in Figure 12.8 by broken lines.

89 Table F provides d-values for sample sizes up to 25. For samples with more than 25 observations, the following normal approximation should be used:

$$d \doteq \frac{1}{2}\left[\frac{1}{2}n(n+1) + 1 - z\sqrt{\frac{n(n+1)(2n+1)}{6}}\right],$$

where z is obtained from Table C according to the confidence coefficient γ (or, in the case of hypothesis testing, according to the significance level α) that we want. For example, we have $n = 30$ observations and should like to carry out a one-sided test (see Section 91) at significance level .05. Table C gives $z = 1.645$ corresponding to $\alpha' = .05$ and

$$d \doteq \frac{1}{2}\left[\frac{30 \times 31}{2} + 1 - 1.645\sqrt{\frac{30 \times 31 \times 61}{6}}\right]$$

$$= \frac{1}{2}[466 - 159.95] \doteq 153.$$

Tests of Hypotheses about Medians

90 In Section 66 of Chapter 8 we pointed out the close relationship between confidence intervals and tests of hypotheses. If η is the median of some population, we can test the hypothesis

$$H: \qquad \eta = \eta_0$$

by finding out whether the hypothetical value η_0 lies in the S- or W-interval. The S-interval discussed in Section 84 is completely general, while the W-interval discussed in Section 86 assumes that the population is symmetric about the population median. Depending on whether or not we are willing to assume symmetry, we then have appropriate methods for testing the hypothesis $\eta = \eta_0$.

We have observed before that a confidence interval contains *all* η-values that are acceptable. But there are times when we are exclusively interested in the acceptability or nonacceptability of one and only one value η_0. Can we test the hypothesis $\eta = \eta_0$ without finding a confidence interval first? We will see that the two confidence procedures are easily converted into test procedures.

Direct test procedures are especially useful when we are interested in one-sided alternatives. The test that rejects the hypothesis $\eta = \eta_0$ when η is not in the S- or W-interval is a test against the two-sided alternative $\eta \neq \eta_0$. It is possible to modify a confidence interval to provide a test against a one-sided alternative. For example, against the alternative $\eta > \eta_0$ we use only the lower bound of the S- or W-interval. If the lower bound is greater than η_0, so that η_0 falls outside the interval as in the accompanying picture, the hypothesis H is rejected in favor of the alternative $\eta > \eta_0$. A direct test procedure is often more convenient.

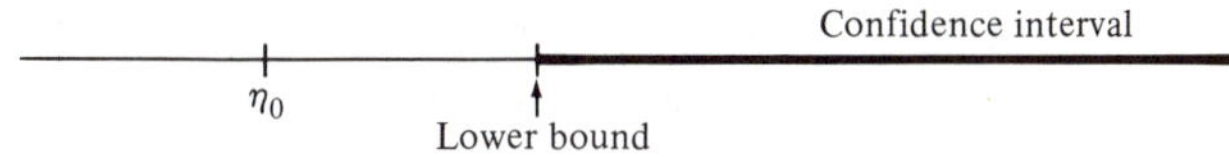

In Section 91 we will discuss the sign test, which corresponds to the S-interval, and in Section 94 we will take up the Wilcoxon signed rank test, which corresponds to the W-interval. A third procedure, the so-called t-test, will be discussed in Chapter 17, where we will make the additional assumption that the underlying population is normally distributed.

91 *The Sign Test*

Since the S-interval for η is bounded by the dth smallest and the dth largest observations, a hypothetical value η_0 must be either smaller than the dth smallest observation or larger than the dth largest observation in order *not* to be acceptable. In the first case, fewer than d observations are smaller than η_0. In the second case, fewer than d observations are larger than η_0. The hypothesis $\eta = \eta_0$ can then be tested in the following way without first finding a confidence interval for η. Let S_- equal the number of observations that are smaller than η_0, and S_+, the number of observations that are greater than η_0. The hypothesis $\eta = \eta_0$ is rejected if either S_- or S_+ is smaller than d. This test is known as the sign test, since it counts how many of the differences $x_i - \eta_0$ are negative and how many are positive.

The procedure just described provides a *two*-sided test, that is, a test against the alternative $\eta \neq \eta_0$. It may appear as if we are using a *one*-tailed critical region consisting of small values of the test statistic against a *two*-sided alternative, contrary to the discussion in Section 62. Actually, there is no contradiction. Since $S_- + S_+ = n$, small values of S_- are associated with large values of S_+ and small values of S_+ are associated with large values of S_-. A completely equivalent test then rejects the null hypothesis provided that S_+ (or S_-) is either sufficiently small *or* sufficiently large. Such a test uses a *two*-tailed critical region. A precise statement of the two-tailed test is given in Problem 25. Our version of the sign test has several advantages. There is no need to make an arbitrary choice between the two test statistics S_- and S_+. It is computationally simpler. Finally, the same table that provides d-values for the S-interval provides critical values for the sign test.

In the case of one-sided alternatives, the correct test statistic is determined by the alternative itself. Since we reject for small values of the test statistic, we use S_- or S_+ depending on which of the two statistics will tend to be small when the alternative is true. Under the alternative $\eta > \eta_0$ we expect few observations to be smaller than η_0 and therefore use S_-. Under the alternative $\eta < \eta_0$, we expect few observations to be greater than η_0 and therefore use S_+.

Table 12.9 summarizes information for the sign test. In each case H is rejected in favor of A provided that the test statistic is *smaller* than d, where d is given in Table E.

The sign test procedure assumes that none of the observations equals η_0. Any observations that are equal to η_0 should be eliminated from the sample and the sample size should be adjusted before carrying out the sign test.

TABLE 12.9

	$H: \quad \eta = \eta_0$	
Alternative A	*Test Statistic*	*Significance Level*
$\eta \neq \eta_0$	smaller of S_- and S_+	$\alpha'' = 1 - \gamma$
$\eta > \eta_0$	S_-	$\alpha' = \frac{1}{2}\alpha'' = \frac{1}{2}(1 - \gamma)$
$\eta < \eta_0$	S_+	$\alpha' = \frac{1}{2}\alpha'' = \frac{1}{2}(1 - \gamma)$

EXAMPLE 12.10 The directors of a bank consider opening a new branch office. They decide that the project is justified only if the median family income in the community is above 10000 dollars. A decision calls for a test of the hypothesis

$$H: \quad \eta \leq 10000$$

against the alternative

$$A: \quad \eta > 10000,$$

where η is the unknown median family income for the community. If an appropriate test rejects H, the directors will feel justified to go ahead with the project.

A random sample of 20 families provides the incomes in Table 12.11. Because of the general asymmetry of income data, the sign test should be used to analyze the data.

TABLE 12.11

9300	50000	15500	7000	12700
10000	10200	9600	15000	35000
11800	7600	21000	19500	17500
6500	14100	9800	12400	8000

According to Table 12.9, the one-sided test should be based on the statistic S_-. We eliminate one observation, which equals the hypothetical value 10000. Among the remaining 19 observations, 7 are smaller than 10000 so that $S_- = 7$. Table E reveals that S_- would have to be smaller than 6 to indicate rejection of H at a conventional level of significance. The sample does not provide convincing evidence that the median family income in the community exceeds 10000 dollars.

92

In Section 63 we pointed out that the computation of the descriptive level associated with a given experimental result may be preferable to an outright test. We want to investigate how to compute descriptive levels associated with sign tests. We have arranged the sign test in such a way that the hypothesis under consideration is rejected only if the appropriate test statistic is sufficiently small. It follows that the descriptive level is always computed as a lower tail probability.

EXAMPLE 12.12 Find the descriptive level associated with Example 12.10. In Example 12.10 we had a one-sided test giving the test statistic $S_- = 7$. The descriptive level is then given by $\delta' = P(S_- \leq 7)$ computed under the assumption that the null hypothesis is true. Table E gives only limited information about the distribution of the sign test statistic. But again the normal distribution provides approximate information. To see how, let us call an observation that is smaller than the hypothetical median a "success" and an observation that is greater, a "failure." If the null hypothesis concerning the population median is correct, success and failure both have probability ½ in view of the definition of the median, so that S_- (as well as S_+) has a binomial distribution with success probability ½. Unless the sample size is quite small, we can then use the normal approximation with

$$\mu = np = \frac{1}{2}n \qquad \text{and} \qquad \sigma = \sqrt{npq} = \frac{1}{2}\sqrt{n}.$$

In our case $n = 19$, so that $\mu = 9.5$ and $\sigma = 2.18$. The descriptive level is given as the area under the corresponding normal distribution to the left of 7, or equivalently, the area under the standard normal curve to the left of $z = (7 - 9.5)/2.18 = -1.15$. From Table B we find that $\delta' = T(-1.15) = T(1.15) = .125$, a result considerably larger than conventional significance levels.

NOTE 12.13* In this and the next two chapters, the accuracy of the computation of the descriptive level can usually be improved by *adding* a continuity correction of ½ to the observed value of the test statistic (see Section 36). In Example 12.12 we would use $z = (7 + \frac{1}{2} - 9.5)/2.18 = -0.92$ giving $\delta' = .179$. Tables of exact binomial probabilities give $P(S_- \leq 7) = .180$. The use of the continuity correction has considerably improved the accuracy of the normal approximation.

93 *Ranks*

Before we can describe the test procedure that corresponds to the *W*-interval, we have to define an important new term. Most nonparametric tests are based on *ranks* rather than actual observations. The rank of a given observation is obtained by arranging all observations according to size and giving the smallest observation the rank 1, the next smallest, the rank 2, and so on. The largest of *n* observations receives the rank *n*.

This procedure assumes that no two observations are equal, or *tied*, as we will say from now on. If ties do occur among the observations, we use average ranks or *midranks*. Thus if we have the five observations

$$6, 8, 8, 11, 15,$$

we would assign the ranks

$$1, 2.5, 2.5, 4, 5,$$

since the two equal observations share the second and third positions when all observations are arranged according to size. Again for the five observations

$$8, 8, 15, 11, 8,$$

we use ranks

$$2, 2, 5, 4, 2,$$

since the three equal observations share the first, second, and third positions when the observations are arranged according to size.

94 *The Wilcoxon Signed Rank Test*

If the sample observations $x_1, \ldots, x_n$ come from a symmetric distribution with center of symmetry η, we can use the W-interval as a confidence interval for η. The W-interval is bounded by the dth smallest and dth largest of the $n(n + 1)/2$ data averages, where d is read from Table F. As we have already pointed out, this confidence interval can be used to test the hypothesis $\eta = \eta_0$. Alternatively, we can use the following direct test procedure. In a manner analogous to that we used with the sign test statistics S_- and S_+, we define two new statistics W_- and W_+ using the data averages instead of the observations. W_- equals the number of data averages that are smaller than the hypothetical population median η_0; W_+ equals the number of data averages that are greater than η_0. The hypothesis $\eta = \eta_0$ is rejected if the smaller of W_- and W_+ is smaller than d. This is a test against the two-sided alternative $\eta \neq \eta_0$. One-sided tests are performed in complete analogy to one-sided sign tests. If η is the center of a symmetric distribution, Table 12.14 summarizes information for the Wilcoxon signed rank test. In each case the hypothesis is rejected provided that the test statistic is *smaller* than d, where d is given in Table F.

TABLE 12.14

H: $\eta = \eta_0$

Alternative A	*Test Statistic*	*Significance Level*
$\eta \neq \eta_0$	smaller of W_- and W_+	$\alpha'' = 1 - \gamma$
$\eta > \eta_0$	W_-	$\alpha' = \frac{1}{2}\alpha'' = \frac{1}{2}(1 - \gamma)$
$\eta < \eta_0$	W_+	$\alpha' = \frac{1}{2}\alpha'' = \frac{1}{2}(1 - \gamma)$

As described, the test seems to require the determination of all $n(n + 1)/2$ data averages. In Section 87 we have seen that the corresponding computations are time consuming even for intermediate sample sizes. Fortunately, there exists a more convenient method for computing W_- and W_+ that does not require the determination of the individual data averages. Instead, we find all differences $x_i - \eta_0$. The absolute values $|x_i - \eta_0|$† are then ranked from 1 to n. W_- equals the sum of the ranks corresponding to negative differences $x_i - \eta_0$; W_+, the sum of the ranks for positive differences $x_i - \eta_0$.

Any observations that equal η_0 are omitted from the sample exactly as in the sign test. Of course, the sample size has to be reduced correspondingly. We always have $W_- + W_+ = 1 + 2 + \cdots + n = n(n + 1)/2$, where n equals the number of nonzero differences $x_i - \eta_0$, so that only one of the two rank sums need be computed from the data. The other rank sum can be obtained by subtraction.

This test is known as Wilcoxon's signed rank or one-sample test.

EXAMPLE 12.15 A machine is supposed to produce wire rods with a diameter of 1 millimeter. To check whether the machine is properly adjusted, 12 rods from the machine's production are selected and measured giving the following results: 1.017, 1.001, 1.008, .995, 1.006, 1.011, 1.009, 1.009, 1.003, .998, .990, and 1.007. Would you say that the machine needs adjustment?

We set up a test of the hypothesis $\eta = 1.000$ against the alternative $\eta \neq 1.000$. If this hypothesis is rejected, machine adjustments are indicated. Industrial measurements of this type are generally symmetrically distributed. We therefore use the Wilcoxon signed rank test. The necessary computations follow:

x_i	$1000\|x_i - 1.000\|$‡	*Rank*
1.017	17	12
1.001	1	1
1.008	8	7
.995	5	4
1.006	6	5
1.011	11	11
1.009	9	8.5
1.009	9	8.5
1.003	3	3
.998	2	2
.990	10	10
1.007	7	6

†The absolute value of a number (indicated by vertical bars) is its numerical value without regard to sign. Thus $|+5| = |-5| = 5$.

‡We multiply by 1000 to avoid decimals. This does not change the ranking.

There are only three negative differences $x_i - \eta_0$, and we find $W_- = 4 + 2 + 10 = 16$. This is the smaller of W_- and W_+, since $W_+ = \frac{1}{2}n(n+1) - W_- = 78 - 16 = 62 > 16$. According to Table F, the descriptive level associated with this result lies between .05 and .09. A checkup on the machine adjustment may be advisable.

As in the case of the sign test, approximate descriptive levels can be obtained with the help of the normal distribution. For the Wilcoxon signed rank test, $\mu = n(n+1)/4$ and $\sigma = \sqrt{n(n+1)(2n+1)/24}$.

EXAMPLE 12.16 When comparing the examination scores of 24 students with the national median for the examination, we find that $W_+ = 78$. What is the descriptive level of this result? The one-sided descriptive level δ' is equal to $P(W_+ \leq 78)$, which can be approximated as the area to the left of 78 under the normal distribution with mean $\mu = 24 \times 25/4 = 150$ and standard deviation $\sigma = \sqrt{24 \times 25 \times 49/24} = 35$. We find that $z = (78 - 150)/35 = -2.06$, so that $\delta' = T(-2.06) = .02$. It would seem that the students are performing below the national average.

95 Occasionally we have analyzed the same set of data by two or more methods. This has been done for illustrative purposes. The student should not conclude that in practical work it is appropriate to analyze a given set of observations by various methods and *then* select the results that appear most satisfactory under the circumstances (for example, select the shortest confidence interval). When planning an experiment, the statistician should decide in advance what method of analysis is to be used on the data resulting from the experiment. Thus in the case of symmetric populations, when both the Wilcoxon signed rank test and the sign test are appropriate, the statistician would very likely decide on the former, since it specifically uses symmetry, unless the greater simplicity of the sign test would appear to outweigh any considerations of greater accuracy.

One important advantage of the Wilcoxon test arises from the fact that the statistic W can take many more different values than the statistic S. As a consequence, the Wilcoxon test has many more possible significance levels than the sign test as we can readily see from a comparison of Tables F and E.

Nonparametric Procedures and Tied Observations

96* In Section 81 we mentioned certain theoretical difficulties associated with nonparametric procedures applied to data measured on a discontinuous scale. The nonparametric methods developed in this chapter provide an opportunity to discuss these theoretical difficulties in greater detail. Nonparametric procedures applied to continuous data are *distribution-free;* when applied to discrete data, they are not distribution-free. By this we mean that the confidence coefficients and significance levels listed in Tables E and F (as well as in similar tables) are exact whatever the sampled population, as long as this population is continuous. For example, the confidence interval for the population median bounded by the second smallest and second largest in a sample of eight observations is .930 whether the sample comes from a population with a distribution like that shown in Figure 11.2 or from one with a normal distribution. The confidence statement is "free" of any distributional assumptions beyond the assumption of continuity. A corresponding statement holds for the significance level of a nonparametric test like the sign test or the Wilcoxon signed rank test. The same is not true when the underlying population is discrete. In that case the exact value of a confidence coefficient or significance level depends on the usually unknown probabilities associated with the various population values. This does not mean that nonparametric methods are inapplicable when analyzing data from discrete populations as is sometimes stated. As we will see, statements that are practically useful are possible even in the discrete case. Before we explain the nature of these statements, it is important to know the reason for the different behavior of nonparametric procedures when applied to continuous and to discrete populations.

Theoretically, when sampling from a continuous population, we can be certain that no ties occur among the observations. However, in practice, no phenomenon produces measurements of a strictly continuous nature. For example, the weight of a person can be any number within a certain range, but due to natural limitations of the accuracy with which a person's weight can be determined, actual weights are stated as 140 or 143 or 143.2 or even 143.18 pounds. These are discrete values. With yearly family incomes, the most accurate statement made (and even the Bureau of Internal Revenue does not require such accuracy) can be only to the nearest cent. On the other hand, measurements like test scores are naturally discrete, but it is often more convenient to treat them as continuous data. Because of this underlying discreteness, we find that ties occur with positive probability. The presence of occasional ties among the observations complicates the application of nonparametric procedures.

What modifications are required then when applying nonparametric methods to discrete data? Let us first consider confidence intervals. In particular, consider the confidence interval for the median η,

$$S_d \leq \eta \leq L_d.$$

We defined the confidence interval as a *closed* interval that included the endpoints as acceptable parameter values. If the underlying population is strictly continuous, the confidence coefficient is not changed if we remove the endpoints from the interval of acceptable parameter values, that is, if we consider the *open* interval

$$S_d < \eta < L_d.$$

The situation is different when the population is discrete. The true confidence coefficient associated with the open interval may then be smaller than that associated with the closed interval. More specifically, the open interval has a true confidence coefficient that is at most equal to the tabulated value γ; the closed interval has a confidence coefficient that is at least equal to the tabulated value γ. Corresponding statements hold for other nonparametric confidence intervals. From a practical point of view, it is always safer to use a closed interval. This is the rule we will follow. As a consequence, true confidence levels will at least equal quoted confidence levels.

As far as confidence intervals are concerned, we use exactly the same procedure whether or not ties are present among the observations. Only the probability statement, that is, the statement involving the confidence coefficient, has to be modified. The situation is rather different when it comes to tests of hypotheses. Here the test procedure itself usually requires modification. The Wilcoxon signed rank test is a good example to illustrate the point.

When there are no ties among the observations, or rather among the absolute differences $|x_i - \eta_o|$, we only need the simple ranks 1, 2, . . . , n when computing the Wilcoxon signed rank test statistic W. The need for midranks arises only when ties are present. Thus the extension of the definition of ranks to include midranks is due to a desire to be able to apply nonparametric tests even when there are ties among the observations. The rule that observations that equal the hypothetical parameter value η_0 be omitted from the analysis in both the sign and Wilcoxon signed rank tests is another modification intended to avoid complications when sampling from discrete populations.

What about the true significance levels for modified test procedures? As we mentioned earlier, when sampling from discrete populations (as evidenced by the presence of ties), we find that the test is no longer distribution-free. However, it turns out that, quite generally, tabulated significance levels furnish upper bounds for the true significance levels. The use of standard tables for modified tests produces *conservative* results. The true descriptive significance level associated with the observed test result is at most as large as the tabulated value α'' for two-sided tests and α' for one-sided tests or the corresponding probabilities obtained from the normal approximation. Actually, for most practical purposes, the difference between the tabulated and true significance levels is negligible, unless a rather large proportion of all observations are tied at just a few values. In such a case a normal approximation incorporating a correction for ties may be helpful.

Since this correction is rarely needed in practice, it is not given in this book.

When sampling from discrete populations, we may not necessarily have complete equivalence between confidence intervals and tests of hypotheses. Thus it may happen that either the lower or upper bound of a confidence interval for the population median η coincides with a hypothetical parameter value η_0. Acceptance or rejection of the hypothesis $\eta = \eta_0$ then depends on whether we use a closed or an open confidence interval. A closed confidence interval would accept the hypothesis $\eta = \eta_0$; an open interval would reject the hypothesis. The use of midranks in the computation of the Wilcoxon signed rank statistic W may be compared to taking a position somewhere between using an open confidence interval and using a closed one. It may happen then that use of a closed confidence interval, as we have recommended, suggests acceptance of the hypothesis $\eta = \eta_0$, while use of the Wilcoxon test based on midranks suggests rejection.

PROBLEMS

Whenever appropriate, state the assumptions about the underlying population distribution that are required to justify your method of solution.

1 Consider the following set of eight observations: 5, 3, -1, 14, 7, -4, 11, 10.

a. Find a confidence interval for the population median using two methods. Use a confidence coefficient close to .94.
b. Test the hypothesis $\eta = 0$ against the alternative $\eta \neq 0$ using two different tests. What is the significance level of each of your tests?
c. Find a point estimate of the population median by two methods.

2 Ten students obtained the following grades on an examination: 72, 95, 79, 83, 93, 80, 91, 74, 70, 86. Test the hypothesis that the median score for this test is 75. Use two different tests.

3 What method would you use to test each of the following hypotheses? In each case state the appropriate test statistic and critical region.

a. *H:* $\eta = 0$ against $\eta < 0$; $n = 30$, $\alpha = .10$
b. *H:* $\eta = 500$ against $\eta \neq 500$; $n = 20$, $\alpha = .01$
c. *H:* $\eta = 100$ against $\eta > 100$; $n = 10$, $\alpha = .05$

4 A food laboratory makes the following seven determinations of the percent of fat content of frankfurters of a certain brand: 19.9, 18.5, 19.8, 19.2, 20.5, 19.1, 19.6.

a. Find a point estimate of the median fat content of frankfurters of this type.
b. Find a confidence interval for the median fat content (γ near .95).

5 A computer installation kept records of the turn-around times of individual jobs. The following are the turn-around times (in minutes) of 15 randomly selected jobs: 6, 4.5, 15, 4, 7, 28, 5.5, 52, 24.5, 12, 5, 6.5, 40, 65, 10.

a. Test the hypothesis that the median turn-around time is $\leq$ 10.
b. Find a confidence interval for the median turn-around time (state your confidence coefficient.)

6 Nine families have the following incomes (dollars per year): 9300, 11200, 17000, 6500, 9100, 8900, 8300, 10200, 7900.

a. Find a point estimate for the median population income.
b. Find a confidence interval for the median income ($\gamma = .96$).

7 Define some student population. Get the weights of 12 *randomly* selected students from this population. (For later use, also write down each student's height.) Find a confidence interval for the median weight of students in your population.

8 The median score on a certain qualifying examination is known to be 75. It is claimed that "special" instruction increases a student's score on the examination. If the scores of 13 students who have had special instruction are available, explain how you would test the correctness of the claim. Be specific. State the hypothesis you would set up, the alternative, the appropriate test statistic and critical region, and any assumptions you would have to make.

9 A testing company finds that 16 tires of a certain make have provided the following number of miles of service:

27900	35100	29800	27700
26700	30700	26900	32400
24800	27400	24900	33300
31600	24300	28300	27600

Do these results support the claim that this kind of tire provides over 30000 miles of service on the average?

10 In each of the following cases, assume that you are testing the hypothesis $\eta = \eta_0$ against the alternative $\eta \neq \eta_0$ and have computed the indicated value of the test statistic. What is the descriptive level of your result?

a. $n = 40$, $S_- = 14$
b. $n = 16$, $W_+ = 30$
c. $n = 16$, $W_- = 33$
d. $n = 100$, $S_+ = 44$
e. $n = 50$, $W_+ = 412$

11 Solve Example 12.15 using the sign test (if appropriate).

12 Solve Example 12.10 using the Wilcoxon signed rank test (if appropriate).

13 Ten light bulbs of a certain type are selected at random from a large shipment of such bulbs. The lifetimes (in hours) of the bulbs are: 220, 2352, 451, 377, 1561, 257, 1329, 111, 876, 525.

a. Estimate the median lifetime of bulbs of this type.
b. Find a confidence interval for the median lifetime.
c. Do the data support the manufacturer's claim that the median lifetime exceeds 1000 hours?
d. Find the descriptive level associated with the hypothesis H: $\eta \leq 1000$.

14 In a simulated traffic reaction study, the following reaction times (in hundredths of a second) were recorded for 12 randomly selected test subjects: 67, 63, 73, 80, 66, 65, 70, 55, 60, 69, 56, 64.

a. Estimate the median reaction time.
b. Find a confidence interval for the median reaction time.
c. Find the descriptive level associated with the hypothesis $\eta = 60$ against $\eta \neq 60$.
d. Should the hypothesis $\eta = 60$ be accepted?

15 The diastolic blood pressure of eight men was found to be: 68, 80, 82, 85, 91, 83, 77, 84.

a. Use the Wilcoxon test to test the hypothesis that the median diastolic blood pressure is 80.
b. Find the W-interval ($\gamma = .96$) for the median diastolic blood pressure.

16 The national median of a certain test is 110. Thirteen students from College C take the test and obtain the following scores: 111, 112, 116, 120, 105, 113, 110, 116, 118, 112, 111, 96, 117. Test the hypothesis that the median score for students from College C is equal to the national median. Use two methods.

17 Twelve rats are fed an experimental diet from birth to age three months. Their weight gains (in grams) are as follows: 77, 62, 66, 65, 74, 70, 68, 71, 72, 67, 71, 70.

a. Find a point estimate for the median weight gain.
b. Find a confidence interval for the median weight gain.
c. Test the hypothesis that the median weight gain is 70.

18 Mrs. M. is checking the gasoline mileage she is getting from her car. For ten tankfuls, she records the following number of miles per gallon: 13.1, 17.2, 15.3, 19.1, 19.0, 20.5, 13.5, 17.2, 13.2, 12.1.

a. Find a point estimate for the median gasoline mileage Mrs. M. gets from her car.
b. What is the descriptive level associated with the hypothesis that the median gasoline mileage is 15 miles per gallon?

19 Consider the data for drug A in Table 13.4.

a. Test the hypothesis that drug A is ineffective in producing additional hours of sleep. Consider a one-sided alternative.
b. Find a point estimate of the median number of additional hours of sleep gained due to drug A.
c. Find a confidence interval for the parameter in part b.

20 Repeat Problem 19 for drug B.

21 A tire company compared tire mileage for male and female drivers. The number of miles each driver obtained from identical sets of tires is as follows:

Male Drivers	20200	23400	22600	27600	16100	21000
	26300	22500	18000	19100	23200	23700
Female Drivers	27400	32400	30100	32200	30600	28900
	29300	24900	27800	34500		

a. Test the hypothesis that the median number of miles a female driver obtains from one set of tires is ≤ 25000.
b. Test the same hypothesis for male drivers.
c. Find confidence intervals for the median number of miles for male and female drivers *separately*.
d. Find a confidence interval for the median number of miles regardless of sex of the driver. To what population of drivers does this confidence interval apply?

22 A newspaper reporter had a damaged car appraised by four insurance company appraisers. The average damage estimate was \$512. The reporter then took the car to ten randomly selected garages to obtain repair estimates. In each case she told the garage that damage to the car was covered by insurance. The various repair estimates were: 559, 574, 618, 489, 727, 754, 670, 596, 651, 469. Set up an appropriate hypothesis and test it. State your alternative.

23 Discuss the nature of type 1 and type 2 errors for the hypothesis testing situation in Example 12.10. What are the consequences of a type 1 error? a type 2 error?

24* The confidence interval for the population median bounded by the second smallest and the second largest observations has confidence coefficient

$$\gamma = 1 - \frac{1+n}{2^{n-1}}.$$

a. Check the appropriate entries in Table E.
b. Derive the formula for γ. (*Hint:* Generalize the derivation in Section 84.)

25* Show that the following test is equivalent to the sign test of the hypothesis $\eta = \eta_0$ against the two-sided alternative $\eta \neq \eta_0$: Reject the hypothesis if either $S_+ \leq d - 1$ or $S_+ \geq n + 1 - d$, where d corresponds to significance level α'' in Table E. (*Hint:* $S_+ + S_- = n$.) Can we use the statistic S_- to test the same hypothesis? How?

26* How can we use the statistic S_+ to test the hypothesis $\eta = \eta_0$ against the alternative $\eta > \eta_0$? (*Hint:* See Problem 25.)

13

COMPARATIVE EXPERIMENTS: PAIRED OBSERVATIONS

Comparative Experiments

97 In practice we often have to select one of several possible ways of performing a certain job. How can we find out which way is best? Or how can we determine whether there is so little difference between the various approaches that it is immaterial which one we choose? In this chapter and the next, we will discuss how to compare two competing methods. In Chapter 15, we will take up the case of more than two methods. Here are a few examples.

Of two (or more) different ways of treating a certain disease, which one produces the best results? Is there any difference in the number of miles per gallon of gasoline that we can get from two (or more) competing makes of cars? Does high-test gasoline give better mileage than regular gasoline in cars that can use either? Is a new, supposedly superior, method suggested to replace an old established method for doing something really superior? For example, is the new mathematics we hear so much about these days really better than what has been taught under the name of mathematics until now? The reader should realize that problems like this last one are usually much more complex than may appear on the surface. We must decide what we mean by new mathematics and old mathematics. Even if we agree on an answer to this question, what do we mean by saying that one method is better than the other? These are nonstatistical problems, but they have to be resolved before we can set up an experiment whose results are to be analyzed by statistical means. It is good to keep in mind that any problem must be formulated properly before statistical theory can be applied successfully to it.

98 *Lotion* X *versus Lotion* Y

Suppose that we want to find out which of two suntan lotions, labeled X and Y for simplicity, provides better protection against sunburn. Let us also assume that eight test subjects have volunteered to participate in an experiment in which they expose their backs to the sun for several hours, protected by suntan lotion. This raises a problem of experimental design. How should the experiment be set up so that it provides the most information about the question under investigation?

We could use lotion X on four of the test subjects and lotion Y on the other four. But a better procedure is to use both lotions on each of the test subjects, one lotion on the right side and the other lotion on the left side. By comparing only right and left sides of the *same* person, we eliminate variability among measurements caused by differences in skin sensitivity. However, one additional precaution is in order. For each person we have to determine, say, by means of the toss of a fair coin, which lotion goes on the left side and which lotion goes on the right side. Since, in general, the right and left sides of a person are not exposed to the sun in exactly the same way, we might introduce an unintentional bias by assigning, say, lotion X always to the right side. *Randomization* introduced by the toss of a coin avoids this and possibly other unsuspected biases. It also provides the basis for the statistical analysis that we are going to carry out.

After the test subjects have exposed their backs to the sun for the prescribed number of hours, we measure the degree of redness on each side. Here are the results in some appropriate unit of measurement:

X	Y	Y	X	Y	X	Y	X
51	46	45	48	53	52	48	62
+5		+3		−1		+14	
X	Y	X	Y	Y	X	X	Y
64	57	51	55	44	55	60	50
+7		−4		+11		+10	

For example, for our first test subject the toss of the coin determined that lotion X should go on the left side, leaving the right side for lotion Y. The accompanying numbers, 51 and 46, are measures of the degree of sunburn, the higher number indicating more severe sunburn. Finally, we have the difference, +5, between the sunburn measurement for the side protected by lotion X and the side protected by lotion Y. For the remaining test subjects, there is similar information.

In each case we subtract the y-measurement from the x-measurement, regardless of whether lotion X was applied to the left side or the right side. Since the differences, x-y, represent the most important information, we list them separately:

13.1 $+5, +3, -1, +14, +7, -4, +11, +10.$

Merely on the basis of inspection of these differences, it is tempting to conclude that lotion Y provides better protection against sunburn than does lotion X. But we should remember our customary approach. Before we conclude that Y is really better, we want to be able to rule out the possibility that our result can be explained simply in terms of chance and that lotions X and Y are in fact equally efficient, or possibly equally inefficient, in preventing sunburn. Thus we set up and test the null hypothesis that there is no difference in the protective ability of the two lotions. Only if our data indicate that this hypothesis should be rejected are we willing to conclude that lotion Y is superior to lotion X. First we must decide how we can translate the hypothesis to be tested into mathematical terms.

Let us have a closer look at our experiment. If there is no difference between the protective properties of the two suntan lotions, the labels X and Y are just that, labels and nothing else. Perhaps the two lotions look different, but as far as sunburn protection is concerned, there is no difference between the two. Then differences between the two measurements of sunburn on the right and left are not caused by differences in protective ability of the two lotions but are due to uncontrolled factors such as changing skin sensitivity and different exposure to the sun. In this case a reversal of the result of the coin toss that determines which lotion goes on the right and which lotion goes on the left would not change the sunburn measurements. But it would change our test results, and a look at the first test subject shows how. We had the results

$$\begin{array}{cc} X & Y \\ 51 & 46 \\ \multicolumn{2}{c}{+5} \end{array}$$

But if the coin had fallen the other way, we would have had the results

$$\begin{array}{cc} Y & X \\ 51 & 46 \\ \multicolumn{2}{c}{-5} \end{array}$$

Thus if there is no difference in the protective ability of our two lotions, a reversal of the coin toss that assigns the two lotions changes a plus sign into a minus sign, and correspondingly, a minus sign into a plus sign. Since a fair coin falls heads or tails with probability $\frac{1}{2}$, it follows that plus and minus signs also have probability $\frac{1}{2}$. As a consequence, if there is no difference in the protective properties of the two lotions, a difference of -5 is just as likely to occur as a difference of $+5$. Of course, the same is true for any other number. In other words, when our hypothesis is correct, differences are symmetrically distributed about zero. In particular, this means that the median difference is zero, $\eta = 0$.

99 In the preceding chapter we discussed two tests of the hypothesis $\eta = \eta_0$, the sign test and the Wilcoxon signed rank test. Because of the observed symmetry both these tests are applicable in the present case with $\eta_0 = 0$. We will illustrate both, although in practice just one would be chosen.

Since there are fewer negative than positive differences, the appropriate test statistic for the sign test is the number of negative differences, of which there are two. Entering Table E with $n = 8$, we find that we cannot reject our hypothesis at any customary significance level.

In the present example it is appropriate to use a two-sided test, since at the start of the experiment we had no idea which lotion might be better. The situation would be different if it were known that lotion Y contains a "miracle" ingredient that is supposed to make it better than lotion X. In that case we would test the hypothesis that lotions X and Y have equal protective ability against the alternative that Y provides better protection. We would then reject the null hypothesis only if the number of negative differences were sufficiently small.

We now turn to the Wilcoxon signed rank test. To compute the test statistic, we need the ranks of the absolute differences:

Difference	+5	+3	−1	+14	+7	−4	+11	+10
Rank of Absolute Difference	4	2	1	8	5	3	7	6

We find that $W_- = 1 + 3 = 4$. According to Table F, we can reject the hypothesis $\eta = 0$ at significance level .055 since W_- is smaller than 5. The Wilcoxon signed rank test would seem to indicate that lotion Y does a better job than lotion X, provided that we are willing to use a significance level as high as .055.

It is instructive to investigate why, in the present example, the Wilcoxon signed rank test suggests rejection of the null hypothesis (at significance level .055) while the sign test does not. Looking at the data, we observe that, in the two cases where lotion X is associated with the less severe sunburn, there is really very little difference between the measurements for lotions X and Y, just 1 and 4 points. On the other hand, in cases where lotion X is associated with more severe burns, we notice such large differences as 10, 11, and 14. As far as the sign test is concerned, any negative difference carries as much weight as any positive difference. In particular, the negative difference −4 counts as much as the positive difference +14. The same is not true of the Wilcoxon signed rank test. Associated with the difference +14 is the rank 8, while associated with the difference −4 is the much smaller rank 3. The sign test only observes that, out of 8 observations, 2 are negative and 6 are positive. The Wilcoxon signed rank test observes, in addition, that the positive differences are considerably larger than the negative differences and therefore have higher ranks.

The reader should not interpret these remarks to imply that the sign test is not a useful test. We intend only to bring out the differences between the two tests. In small samples the sign test generally conveys little information. But in large samples it often provides all the information that is needed to reach a sound conclusion. In addition, what the sign test lacks in sensitivity, it makes up in versatility. There are situations when only the sign test can be used. For example, suppose that in the suntan lotion problem we had only statements from each of the test subjects as to which side felt more sensitive from exposure to the sun rather than actual measurements of degree of redness. As before, we can mark down a plus sign if the more sensitive side was protected by lotion X and a minus sign if it was protected by lotion Y and then carry out the sign test. The sign test does not really require numerical information. Qualitative judgments are quite sufficient.

In many consumer preference studies, qualitative comparisons are the only ones obtainable. If you are asked to compare two different kinds of ice cream, you can presumably tell which kind you prefer, but you may not be able to give a meaningful numerical value for your degree of preference. In such cases the sign test offers the only possible method of analysis. In the suntan lotion example, both tests are applicable. However, in view of the small number of observations, most statisticians would presumably choose the Wilcoxon signed rank test.

The Analysis of Paired Observations

100 We are now ready to formulate our problem and its solution in general terms. But first we want to introduce some standard terminology. The problem of comparing two or more methods occurs so frequently that statisticians have developed a specialized terminology for it. Rather than speak of methods, they speak of treatments. The hypothesis we are testing states that the two or more treatments we are comparing have identical effects. Only if this hypothesis is rejected, are we willing to concede that differences exist.

In this chapter we assume that we have two treatments and that we can obtain n pairs of observations, where one observation in each pair is from treatment 1 and the other from treatment 2. It is convenient to write these pairs of observations as $(x_1, y_1), \ldots, (x_n, y_n)$. Our analysis is based on the n differences

$$d_1 = x_1 - y_1$$
$$\vdots$$
$$d_n = x_n - y_n.$$

Our discussion for the suntan lotion experiment shows that when the hypothesis of identical treatment effects is correct, the differences $d_1, \ldots, d_n$ have symmetric distributions with center of symmetry η at zero. On the other hand, when the two treatments have different effects, the differences $d_1, \ldots, d_n$ may be assumed to have distributions with median different from zero. We have reduced our problem to one of testing the hypothesis $\eta = 0$ on the basis of n observed differences $d_1, \ldots, d_n$ against the two-sided alternative, $\eta \neq 0$, or possibly the one-sided alternatives, $\eta < 0$ or $\eta > 0$.

In Chapter 12 we saw that this hypothesis can be tested by either the sign test or the Wilcoxon signed rank test. Since the hypothetical value η_0 equals zero, the test statistics are particularly simple. Thus for the sign test, $S_- = \#$ (negative differences) and $S_+ = \#$ (positive differences). For the Wilcoxon signed rank test, the absolute differences $|d_1|, \ldots, |d_n|$ are ranked from 1 to n, then W_- is set equal to the sum of ranks corresponding to negative d's and W_+ is set equal to the sum of ranks corresponding to positive d's. For both the sign and the Wilcoxon signed rank tests, any pair of observations with $x_i = y_i$ is omitted from consideration (that is, differences of 0 are omitted) and the sample size is reduced correspondingly.

EXAMPLE 13.2 Suppose that two methods of teaching reading are to be compared. It is claimed that method 2 is more effective than method 1, where the comparison is based on scores on some test measuring reading ability. The following experiment is conducted. Twenty first graders are divided into ten groups of two. The two pupils in each group are selected so as to be as similar as possible with respect to previous reading experience, IQ, motivation, and other factors that may have a bearing on a student's ability to learn to read. Then in a random fashion one student in each group is assigned to reading method 1, the other to reading method 2. At the end of one year's instruction all 20 pupils take the same test. Suppose that the results (the x-value in each pair is the score of the student taught by method 1) are as recorded in Table 13.3.

TABLE 13.3

Group	1	2	3	4	5	6	7	8	9	10
x	82	81	69	86	70	66	76	63	62	74
y	68	99	67	97	75	86	92	75	84	66
$d = x - y$	14	−18	2	−11	−5	−20	−16	−12	−22	8
Rank	6	8	1	4	2	9	7	5	10	3

For illustrative purposes, we again apply both the sign test and the Wilcoxon signed rank test. Since it has been claimed that method 2 (y-scores) is better, we set up the hypothesis that method 1 is at least as good as method 2. Thus we have a test of the hypothesis $\eta \geq 0$ against the alternative $\eta < 0$. The sign test statistic is $S_+ = 3$, which is not significant at levels given in Table E. Table A of the binomial distribution with $p = \frac{1}{2}$ provides the exact descriptive level: $P(S_+ \leq 3) = b(3) + b(2) + b(1) + b(0) = .117 + .044 + .010 + .001 = .172$.

For the Wilcoxon signed rank test, we find that $W_+ = 6 + 1 + 3 = 10$. According to Table F, the one-sided descriptive level equals $P(W_+ \leq 10) = P(W_+ < 11) = .042$. At the .05 level of significance, we would agree with the claim that method 2 is more effective than method 1.

101 *Confidence Intervals*

In this chapter we have concentrated on the discussion of tests of hypotheses. The two procedures for finding confidence intervals for a population median discussed in Chapter 12 can be applied to the differences $d_i = x_i - y_i$ to find a confidence interval for the median of all possible differences. Indeed, in Problem 1 of Chapter 12 the reader was asked to find a confidence interval using as data the eight differences for the suntan lotion experiment. The resulting interval represents a confidence interval for the median difference in protection between lotions X and Y.

Example 13.2 was formulated in such a way that it required a yes or no answer to the question: Is method 2 superior to method 1? This formulation of the problem is appropriate if a decision has to be made whether method 2 should be used exclusively in the future. On the other hand, when no such decision is contemplated, it may be much more informative to find a confidence interval for the median difference of scores to be achieved by the two methods.

PROBLEMS

1 To compare the speed of two types of desk calculators in performing statistical calculations, a trained operator performed six standard computations on each of the calculators. The following table gives the number of seconds it took to perform the various computations:

Computation	*Calculator 1*	*Calculator 2*
1	25	23
2	62	75
3	46	56
4	123	167
5	89	95
6	365	429

Is there a difference in the operating speed of the two calculators?

2 To evaluate the effectiveness of a tranquilizer drug, a doctor assigns "anxiety scores" to nine patients before and after administration of the drug (high scores indicate greater anxiety):

Patient	1	2	3	4	5	6	7	8	9
Before	23	18	18	16	22	28	15	21	20
After	20	12	14	18	21	20	20	24	13

Would you say that the drug lessens anxiety (as measured by the anxiety score)?

3 Two different coffee makers are being studied to see if there is any difference in the length of time it takes to brew a pot of coffee. Since initial water temperature varies from day to day but is approximately the same on a given day for each pot, the experiment is run with paired observations. The following table gives the length of time (in minutes) it took to brew coffee on the various days:

Day	1	2	3	4	5	6	7	8	9	10
Make A	9.9	9.7	8.8	9.8	6.8	7.3	12.0	11.0	12.2	14.1
Make B	14.3	11.5	8.8	13.5	8.2	10.1	10.1	11.9	13.3	17.3

Does the brewing time differ for the two coffee makers?

4 Water samples from eight sites on a river before and two years after an antipollution program was started gave the following results. The numbers represent scores for a combined pollution measure, higher scores indicating greater pollution:

Site	1	2	3	4	5	6	7	8
Initial Scores	88.4	68.9	100.5	81.4	96.3	73.7	65.1	72.1
Scores After 2 Years	87.1	69.1	91.1	75.6	96.9	69.2	66.3	68.3

We are interested in learning whether the antipollution program has been effective in reducing pollution.

5 Consider the data in Table 15.6 comparing gasoline mileage for different makes of cars. Consider only makes *C* and *F*. Let η be the median difference in gasoline mileage between make *C* and make *F*.

a. Test the hypothesis that $\eta = 0$.
b. Find a confidence interval for η.
c. Find a point estimate of η.

6 The pulse rates of 12 students before and after exercise were as follows:

Before	62	65	71	58	75	68	60	72	75	91	59	68
After	80	79	96	87	87	91	90	92	96	89	81	84

Find confidence intervals (by two different methods) for

a. the median pulse rate before exercise;
b. the median increase in pulse rate due to exercise.

7 To compare the effects of two sleep-inducing drugs, the following experiment was conducted. On different nights each of ten test subjects received either drug A or drug B. The table gives the additional hours of sleep (compared to a test subject's customary number of hours of sleep) for the two drugs.

a. Is there any difference in the number of additional hours of sleep for the two drugs?
b. Find a point estimate for the median difference of additional hours of sleep.
c. Find a confidence interval for the parameter in part b.

Patient	*Drug A*	*Drug B*
1	0.7	1.9
2	−1.6	0.8
3	−0.2	1.1
4	−1.2	0.1
5	−0.1	−0.1
6	3.4	4.4
7	3.7	5.5
8	0.8	1.6
9	0.0	4.6
10	2.0	3.4

8 To investigate the influence of type of tire on gasoline consumption, the following experiment was conducted. Eight different cars equipped with radial tires were driven over a standard course. Cars were then equipped with regular tires and driven over the same course. The gasoline mileage (miles per gallon) obtained in each case was as follows:

Car	1	2	3	4	5	6	7	8
Radial Tires	26.5	14.3	12.7	20.2	15.1	16.9	23.4	16.4
Regular Tires	25.8	14.5	12.1	19.9	15.1	15.8	23.0	16.0

a. Find a point estimate for the median differences in gasoline mileage for the two types of tires.
b. Find a confidence interval for the median difference in gasoline mileage.

9 Fifty persons were asked which of two brands of soft drinks they preferred.

a. If 18 stated that they preferred brand X, 26 stated that they preferred brand Y, and 6 stated that they did not like either, would it be fair to say that brand y is preferred to brand X? Explain your reasoning.
b. If the numbers were 19, 31, and 0, respectively, how would you answer the question in part a?

10 An advertisement for a certain brand of electric shaver reported that, in 225 "split-face" tests, 56 percent of the time their electric razor shaved at least as close as a barber's straight razor. A footnote, in small print, gave the following additional information: closer, 36 percent; as close, 20 percent. Comment.

14

COMPARATIVE EXPERIMENTS: TWO INDEPENDENT SAMPLES

102 In the preceding chapter we analyzed comparative experiments in which measurements occur in pairs. For instance, each test subject in the suntan lotion experiment provided a measurement pair for lotions X and Y. Our analysis was based on the differences $x_i - y_i$. There are situations when it is not possible to use two different treatments for the same test subject. If we want to compare the effectiveness of two different medical treatments, it is generally impossible to apply both to the same patient. In determining which of two methods of teaching first graders how to read is more successful, we cannot use both teaching methods on the same child. We got around the difficulty in this particular case by dividing the participating children into groups of two in such a way that the two children forming a pair were as homogeneous as possible with respect to factors thought to have some bearing on a child's ability to learn to read. By assigning one child in each group to method 1 and the other child to method 2, we obtained paired observations that could be analyzed according to the methods of Chapter 13. However, the success of this approach depends on how successful we are in forming homogeneous pairs. If no information is available or if no homogeneous pairs can be found, it is better not to attempt to use pairs as the basis of our experiment.

As an alternative, we divide the available test subjects in two groups. We use method 1 on the test subjects in the first group and method 2 on the test subjects in the second group. In this chapter we discuss methods for analyzing data resulting from such experiments. Some general remarks are in order before we do so.

It is usually best to keep the two experimental groups as nearly equal in number as possible. However, it is not always possible—or practical—to have exactly the same number of test subjects in each group. For this reason, we want a method of analysis that is applicable whether the groups are of equal size or not.

In the suntan lotion experiment, we randomly assigned lotions X and Y to the right or left sides of the backs of the test subjects to avoid possible biases. For the same reason, it is desirable to divide the available test subjects into two experimental groups in a random fashion. Many experiments have been spoiled from the beginning by a nonrandom assignment of test subjects. If a subsequent analysis reveals group differences, there is always the possibility that the observed differences do not reflect differences between methods 1 and 2. They may be due to nonrandom assignments. In the reading experiment, suppose that one school system uses method 1 and another school system uses method 2. At the close of first grade, the children in both school systems are given the same test, and the children in group 2 get significantly higher scores than those in group 1. Does this mean that method 2 is superior to method 1? Not necessarily. The kindergarten preparation of the children in group 2 may have laid much more emphasis on preparation for reading than in group 1. There are other possible reasons. Only by assigning the available children to the two reading classes at random is it possible to avoid biases of this type.

Of course, experimental conditions may not permit an effective randomization of observations for the two treatments. If this is the case, it is important for the experimenter to make certain that any potential experimental differences are due to treatment differences rather than extraneous factors.

We turn now to the problem of how to analyze two (independent) sets of observations from a comparative experiment. Let us look at some data. Here are the scores that two different groups of pupils obtained on an examination. (We will find it convenient to refer to the first set as x-scores and to the second set as y-scores.)

TABLE 14.1

Group 1 (x-scores)	72	79	93	91	70	95	82	80	74	86	
Group 2 (y-scores)	78	66	65	84	69	73	71	75	68	90	76

We assume that pupils in groups 1 and 2 were taught by two different methods, and we want to find out if the two methods of teaching are equally effective.

Even though there are only 21 observations in all, it is difficult to get a clear picture of what is going on simply by looking at these numbers. A graph is more enlightening. In Figure 14.2 the 21 scores are

FIGURE 14.2

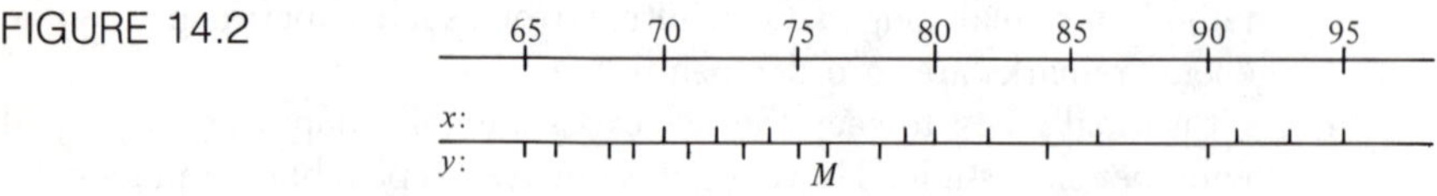

plotted along a horizontal axis, x-scores on top, y-scores below. On the whole it would seem that pupils in group 1 have been doing better than pupils in group 2. But we should certainly know by now that general impressions are insufficient to establish the superiority of teaching method 1 over teaching method 2. We need formal analysis to back up, or correct, our intuitive feeling. In particular, we want to be able to rule out the possibility that chance alone has brought about the configuration of results we have observed and that a repetition of the experiment could reverse the picture.

A formal comparison of x- and y-scores can take many different forms, resulting in different tests of the null hypothesis that there is no difference in the effectiveness of the two teaching methods. In this chapter we look at two such tests. A third test assuming normally distributed populations is discussed in Chapter 18.

The Median Test

103 We start by finding the median M of all x- and y-scores. In our example, $M = 76$. We then divide the 21 scores into two groups. The first group consists of scores that are greater than the median (in symbols, $> M$). The second group consists of the remaining scores (in symbols, $\leq M$). Finally, we find out how many x-scores and how many y-scores belong in each of these two groups. A two-way table like Table 14.3 is most convenient for tabulating the result.

TABLE 14.3

	x-scores	*y-scores*	*Totals*
$\leq M$	3	8	11
$> M$	7	3	10
	10	11	21

When the null hypothesis is correct, whether an observation falls above M or not is unrelated to whether it is an x- or a y-observation. To test this proposition, we apply the chi-square test of Section 75 to Table 14.3. Easy computations give

$$\chi^2 = \frac{(3 \times 3 - 7 \times 8)^2\, 21}{11 \times 10 \times 10 \times 11} = 3.83.$$

For one degree of freedom, Table D shows that the descriptive level associated with χ^2 is just about .05. There is some evidence that the number of x-scores greater than M is larger than can reasonably be expected under the null hypothesis. This test is known as the *median test.*

Like the sign test, the median test counts how many observations are above or below the median. It does not pay attention to how much or how little a given observation differs from the median. The Wilcoxon test, to be discussed next, makes use of the latter kind of information. The chief recommendation of the median test is its simplicity both conceptually and computationally. Like the sign test, it has its limitations, particularly when based on few observations. On the other hand, it is quick and often does the job with a minimum of effort.

The Wilcoxon Two-sample Test

104 The Wilcoxon two-sample test, or Wilcoxon-Mann-Whitney test as it is sometimes called, compares every x-score with every y-score counting the number U_x of times x-scores surpass y-scores, or alternatively, the number U_y of times y-scores surpass x-scores. From Figure 14.2 we see that the lowest x-score surpasses 4 y-scores, the next lowest x-score surpasses 5 y-scores, and so on. In this way we find that

$$U_x = 4 + 5 + 6 + 9 + 9 + 9 + 10 + 11 + 11 + 11 = 85.$$

The computation of U_y is even faster,

$$U_y = 0 + 0 + 0 + 0 + 1 + 2 + 3 + 3 + 3 + 6 + 7 = 25.$$

When there is no difference in the effectiveness of the two teaching methods, we would expect x- and y-scores to be similar. This similarity in scores results in similar values for U_x and U_y. On the other hand, when y-scores tend to be larger than x-scores, then U_x should be small; and if x-scores tend to be larger than y-scores, then U_y should be small. This discrepancy in scores suggests rejection of the null hypothesis when the smaller of U_x and U_y is sufficiently small. As we will see, U_y is sufficiently small for our data to suggest rejection of the null hypothesis at the .05 level.

105 We will discuss the Wilcoxon two-sample test in general. Assume that we have two independent sets of observations, an x-set consisting of m observations

$$x_1, \ldots, x_m$$

and a y-set consisting of n observations

$$y_1, \ldots, y_n.$$

In the preceding example, these are the two sets of test scores and, of course, $m = 10$ and $n = 11$. The hypothesis we want to test states that these two sets of observations have come from one and the same population. We are willing to believe that methods 1 and 2 produce different results only if our data indicate that the hypothesis we have just stated should be rejected.

For the sake of our present discussion, we assume that there are no ties among the observations, and, in particular, that no x-observation equals a y-observation. We will discuss required modifications when ties occur in Section 107.

As in the preceding example, we let

$$U_x = \text{number of times } x\text{-observations are larger than } y\text{-observations,}$$
$$U_y = \text{number of times } y\text{-observations are larger than } x\text{-observations.}$$

To compute U_x and U_y, we have to compare every x-observation with every y-observation, mn comparisons in all. It then follows that $U_x + U_y = mn$, so that only one of the two quantities has to be computed. The other can always be obtained by subtraction. In the earlier example inspection shows that the computation of U_y requires much less time than the computation of U_x. One advantage of computing both quantities from the data is that we can check the correctness of our computations by checking that $U_x + U_y = mn$.

It is not necessary to arrange observations according to size as in Figure 14.2 to compute U_x or U_y. Either statistic can be computed just as easily from the unordered observations in Table 14.1. Thus if we have decided to compute U_y, we start with the first y-observation, 78, and note that it surpasses three x-observations, namely, 72, 70, and 74. The second y-observation surpasses zero x-observations. Continuing in this way, we find that

$$U_y = 3 + 0 + 0 + 6 + 0 + 2 + 1 + 3 + 0 + 7 + 3 = 25$$

as before. The Wilcoxon test rejects the hypothesis that the x- and y-samples have come from the same population if the smaller of U_x and U_y is *smaller* than d, where d is listed in Table G. This test has significance level α'' and is a test against the two-sided alternative which states that x-observations tend to be greater than y-observations or, vice versa, that y-observations tend to be greater than x-observations.

Against one-sided alternatives, the correct test statistic is determined by the alternative. If, according to the alternative, y-observations tend to be smaller than x-observations, then fewer y-observations will surpass x-observations when the alternative is true than when the null hypothesis is true, and we use U_y as the test statistic. Correspondingly, against the alternative that x-observations tend to be smaller than y-observations, we use U_x as the test statistic.

Table 14.4 summarizes information for the Wilcoxon two-sample test. In each case H is rejected in favor of A if the test statistic is *smaller* than d, where d is given in Table G.

Table G lists d-values for sample sizes m and n between 3 and 12. The appropriate entries are found at the intersection of the row corresponding to the number of observations in the larger sample and the

TABLE 14.4 *H:* *x*- and *y*-observations come from identical populations

Alternative A	*Test Statistic*	*Significance Level*
x- and *y*-observations come from different populations.	smaller of U_x and U_y	$\alpha'' = 1 - \gamma$
y-observations tend to be smaller than *x*-observations.	U_y	$\alpha' = \frac{1}{2}\alpha'' = \frac{1}{2}(1 - \gamma)$
x-observations tend to be smaller than *y*-observations.	U_x	$\alpha' = \frac{1}{2}\alpha'' = \frac{1}{2}(1 - \gamma)$

column corresponding to the number of observations in the smaller sample. Thus for the example in Section 104, we find sample size 11 in the left-hand margin and sample size 10 in the top row. Table G tells us that the null hypothesis should be rejected at significance level $\alpha'' = .051$ provided that the smaller of U_x and U_y is smaller than $d = 28$. Since $U_y = 25$, rejection of the hypothesis that there is no difference in the effectiveness of the two teaching methods is indicated at the .05 level.

If at least one sample contains more than 12 observations, the following normal approximation should be used to get *d*-values,

$$d \doteq \frac{1}{2}[mn + 1 - z\sqrt{mn(m + n + 1)/3}],$$

where z is obtained from Table C according to the significance level α (or, in case of confidence intervals, according to the confidence coefficient γ) that we want to achieve. For example, we have two samples with 15 and 20 observations, respectively, and should like to carry out a two-sided test at significance level .05. Table C gives $z = 1.96$ corresponding to $\alpha'' = .05$ and

$$d \doteq \frac{1}{2}[15 \times 20 + 1 - 1.96\sqrt{(15 \times 20 \times 36)/3}]$$

$$\doteq \frac{1}{2}[301 - 117.6] = 91.7.$$

We would then use $d = 92$.

106* Exact Distribution of Test Statistic

In general, we do not derive the sampling distributions that underlie given test and estimation procedures. One advantage of the nonparametric approach is the ease with which such sampling distributions can be derived, at least for small samples. We will illustrate this point using the Wilcoxon test for sample sizes $m = 2$ and $n = 3$. We let U stand for the smaller of U_x and U_y.

Since the values of the statistics U_x and U_y depend only on the position of *x*-observations relative to *y*-observations, we need only enumerate all possible arrangements of two x's and three y's and compute the appropriate value of U for each. Thus we may write

$$xxyyy$$

to indicate that both x-observations are smaller than all three y-observations. For this arrangement, it is clear that $U_x = 0$, and $U_y = 6$, and therefore $U = 0$. In all there are ten different arrangements of two x's and three y's. Table 14.5 lists all ten arrangements together with

TABLE 14.5 **Values of U; $m = 2$, $n = 3$**

Arrangement	U_x	U_y	U
xxyyy	0	6	0
xyxyy	1	5	1
xyyxy	2	4	2
xyyyx	3	3	3
yxxyy	2	4	2
yxyxy	3	3	3
yxyyx	4	2	2
yyxxy	4	2	2
yyxyx	5	1	1
yyyxx	6	0	0

the corresponding values of U_x, U_y, and U. Since the null hypothesis states that the x- and y-observations come from one and the same population, so that x's and y's are meaningless labels, all ten arrangements have the same probability, namely, $1/10$. We then find the distribution for U given in Table 14.6. In particular, the test that rejects the null hypothesis when $U = 0$ has significance level .20. The corresponding entry in Table G would read:

$$d = 1 \qquad \gamma = .80 \qquad \alpha'' = .20 \qquad \alpha' = .10.$$

TABLE 14.6 **Distribution of U; $m = 2$, $n = 3$**

$P(U = 0) = 2/10$
$P(U = 1) = 2/10$
$P(U = 2) = 4/10$
$P(U = 3) = 2/10$

107 *Tied Observations*

Our preceding discussion of the U-test is based on the assumption that there are no ties among the observations. Since ties do occur in practice, we will now indicate how to deal with them. First of all, ties among x- and y-observations require a redefinition of U_x and U_y. According to our earlier instructions, to find the value of U_x, we compare every x-observation with every y-observation, count 1 whenever the x-observation is greater than the y-observation, and count 0 when it is less. We now add the following rule. Whenever an x-observation equals a y-observation, we count $\frac{1}{2}$. A corresponding change applies to the computation of U_y. This additional rule does not change the relationship $U_x + U_y = mn$, since a tie between an x- and a y-observation adds $\frac{1}{2}$ to both U_x and U_y. Even in the presence of ties, it is sufficient to compute one of the two statistics and find the other by subtraction.

EXAMPLE 14.7 A car owner wants to know whether to believe the advertising claim that high-test gasoline improves gas mileage and so conducts the following experiment. The car owner observes repeatedly the number of miles driven with ten gallons of gasoline. Here are the results:

High-test gasoline (x-scores)	175	139	186	228	211	182	185
Regular gasoline (y-scores)	186	206	149	213	156	181	180

To get an answer to the question, the car owner tests the hypothesis that all 14 observations have come from the same population against the alternative that high-test gasoline produces better gas mileage than regular gasoline. The appropriate test statistic is U_y. Using the computation procedure of Section 105, we find that $U_y = 4.5 + 5 + 1 + 6 + 1 + 2 + 2 = 21.5$. (The entry 4.5 indicates that the first y-observation surpasses four x-observations and is equal to a fifth one.) According to Table G, this result is highly insignificant. There is no evidence that high-test gasoline provides better gas mileage than regular gasoline.

108 *Descriptive Levels*

As usual, approximate descriptive levels for the Wilcoxon two-sample test are obtained with the help of the normal distribution, unless either sample size is quite small. The appropriate mean and standard deviation are $\mu = \frac{1}{2}mn$ and $\sigma = \sqrt{mn(m + n + 1)/12}$.

EXAMPLE 14.8 Find the descriptive level associated with Example 14.7. In Example 14.7 we had $m = n = 7$ and $U_y = 21.5$. The one-sided descriptive level is given by $\delta' = P(U_y \leq 21.5)$, which is approximated as the area to the left of 21.5 under the normal distribution with mean $\mu = 7 \times 7/2 = 24.5$ and standard deviation $\sigma = \sqrt{(7 \times 7 \times 15)/12} = 7.83$ or, equivalently, as the area under the standard normal curve to the left of $z = (21.5 - 24.5)/7.83 = -0.38$. Thus $\delta' \doteq T(-0.38) = .352$. This result gives numerical support to our earlier statement that, as far as gasoline mileage is concerned, there does not seem to be any difference between regular and high-test gasoline.

109 *Alternative Form of Wilcoxon Two-Sample Test*

Wilcoxon actually proposed a statistic other than U for carrying out the test that bears his name. To find this other statistic, we arrange the $(m + n)$ x- and y-observations according to size and associate with each observation its rank as in Figure 14.9. Let R_x be the sum of the ranks corresponding to x-observations and let R_y be the sum of the ranks corresponding to y-observations. Wilcoxon suggested using R_y as the test statistic and rejecting the null hypothesis (against a two-sided alternative) whenever R_y is sufficiently large or sufficiently small. Many statistics texts refer to the test based on R_y as the Wilcoxon test and to the test based on U_x or U_y as the Mann-Whitney test.

FIGURE 14.9

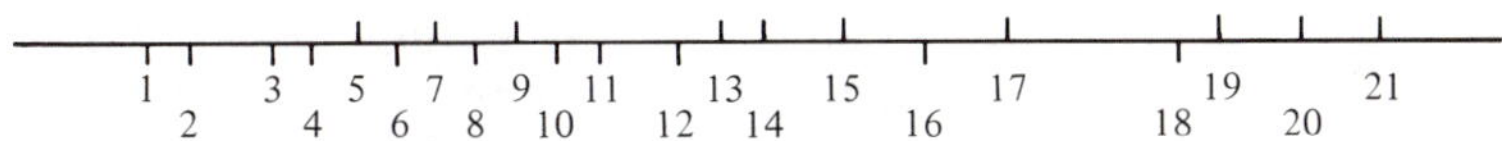

We want to show now that the four quantities U_x, U_y, R_x, and R_y are closely related. If we know one of the four, we can easily compute the other three. We know already that $U_x + U_y = mn$, allowing us to compute one from the other. From the definitions of R_x and R_y, we have

$$R_x + R_y = 1 + \cdot \cdot \cdot + (m + n) = \frac{1}{2}(m + n)(m + n + 1),$$

allowing us to compute one rank sum from the other. The relationship between R_x and U_x and between R_y and U_y is slightly more complicated. A mathematical analysis of the computation of ranks shows that

$$R_x = U_x + (1 + 2 + \cdot \cdot \cdot + m) = U_x + \frac{1}{2}m(m + 1)$$

and, similarly, that

$$R_y = U_y + \frac{1}{2}n(n + 1).$$

In our example, $n = 11$, $U_y = 25$, and $R_y = 91$, which satisfy the above equality.

EXAMPLE 14.10 Compute R_y for the data in Example 14.7 and check the value of U_y. The following arrangement of the observations gives us the value of R_y:

Observation	139	149	156	175	180	181	182	185
x or y	*x*	*y*	*y*	*x*	*y*	*y*	*x*	*x*
Rank	*1*	*2*	*3*	*4*	*5*	*6*	*7*	*8*

Observation	186	186	206	211	213	228
x or y	*x*	*y*	*y*	*x*	*y*	*x*
Rank	9.5	9.5	11	12	13	14

(Since an x- and a y-observation are tied for ninth and tenth place, both receive the midrank 9.5.) We then find $R_y = 2 + 3 + 5 + 6 + 9.5 + 11 + 13 = 49.5$ and $U_y = R_y - \frac{1}{2}n(n + 1) = 49.5 - 28 = 21.5$ as before.

There are two reasons for introducing the rank sums R_x and R_y. As we have just seen, they provide an alternative way for computing the test statistics U_x and U_y. More importantly, however, in Chapter 15 we discuss how to compare three or more treatments. It is simpler to generalize from the Wilcoxon test using rank sums R_x and R_y than from the test based on the statistics U_x and U_y.

Confidence Intervals for a Shift Parameter

110 In this chapter as in Chapter 13 we have primarily concerned ourselves with tests of hypotheses. However, experience shows that observations from two comparable, though different, treatments often follow distributions that are identical in shape but shifted relative to one another as in Figure 14.11. If this seems a reasonable model, the experimenter

FIGURE 14.11

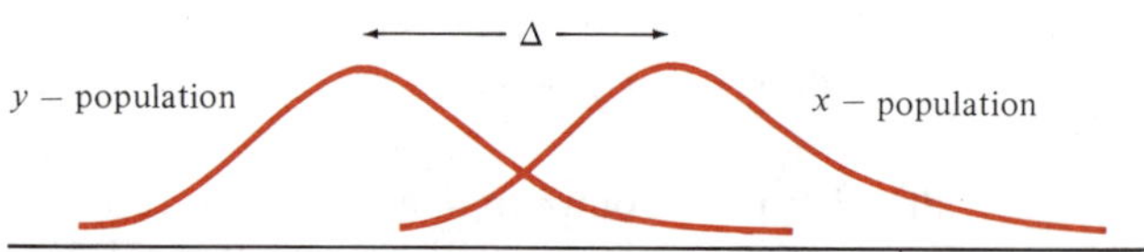

will be interested to find a confidence interval for the amount of shift Δ of the x-population relative to the y-population. Such a confidence interval can be readily determined. We find all possible differences between an x-observation and a y-observation: $x_1 - y_1, x_1 - y_2, \ldots, x_i - y_j, \ldots, x_m - y_n$. There are mn such differences, each of which may be considered an estimate of Δ. A confidence interval for Δ is bounded by the dth smallest and dth largest of all these "estimates," where d is read from Table G.

There is a close relationship between this confidence interval and the Wilcoxon test discussed earlier in this chapter. If the shift parameter Δ in Figure 14.11 is zero, the x- and y-observations come from the same population and the null hypothesis tested by the Wilcoxon test is satisfied. Our confidence interval contains the parameter value $\Delta = 0$ if and only if the Wilcoxon test accepts the hypothesis that the x- and y-observations have come from the same population. As on earlier occasions, the confidence interval can be used in place of a test of a hypothesis.

As an illustration, let us find a confidence interval for the shift parameter Δ using the data in Table 14.1. For $m = 10$, $n = 11$, we find from Table G that the value $d = 20$ corresponds to a confidence interval with confidence coefficient .99. For such a confidence interval, we need the 20th smallest and the 20th largest of the 110 possible differences $x_i - y_j$. Using either Table 14.12 or the graphical method of Section 112, we find that the 20th smallest difference equals -3 and the 20th largest, $+20$. Thus we have the confidence interval

$$-3 \leq \Delta \leq 20.$$

The median score of students taught by method 1 may be as much as 20 points higher than that of students taught by method 2. But it may also be as much as 3 points lower.

We note that the confidence interval does contain the value $\Delta = 0$. This is no contradiction to our earlier test result. The confidence interval provides a test with significance level $\alpha'' = 1 - \gamma = 1 - .99 = .01$. The earlier test indicated that the null hypothesis could be rejected at significance level .05, but not at the level .01. Indeed, the descriptive level is about .04.

The mn differences $x_i - y_j$ are most conveniently exhibited in a two-way table like Table 14.12 in which the entries correspond to increasing x-observations and decreasing y-observations. For example,

TABLE 14.12

y \ x	70	72	74	79	80	82	86	91	93	95
90	−20	−18	−16	−11	−10	−8	−4	1	3	5
84	−14	−12	−10	−5	−4	−2	2	7	9	11
78	−8	−6	−4	1	2	4	8	13	15	17
76	−6	−4	−2	3	4	6	10	15	17	19
75	−5	−3	−1	4	5	7	11	16	18	20
73	−3	−1	1	6	7	9	13	18	20	22
71	−1	1	3	8	9	11	15	20	22	24
69	1	3	5	10	11	13	17	22	24	26
68	2	4	6	11	12	14	18	23	25	27
66	4	6	8	13	14	16	20	25	27	29
65	5	7	9	14	15	17	21	26	28	30

the entry 10 at the intersection of the seventh column and fourth row represents the difference between the seventh *smallest* x-observation, 86, and fourth *largest* y-observation, 76.

111*

The following scheme saves a considerable amount of computational effort. To simplify the presentation, we use only the four largest x-observations and the three smallest y-observations in Table 14.1. Our first step consists in arranging the x-observations from the smallest to the largest and determining the successive differences which we will call x-increments,

x-observations	86	91	93	95
x-increments		5	2	2

Next we arrange the y-observations from the largest to the smallest. The resulting differences will be called y-decrements,

y-observations	68	66	65
y-decrements		2	1

We obtain our first row of $(x - y)$-differences by subtracting the largest y-observation from the smallest x-observation and then successively adding the x-increments:

$$86 - 68 = 18 \qquad 18 + 5 = 23 \qquad 23 + 2 = 25 \qquad 25 + 2 = 27.$$

To get the second row of differences, we add the first y-decrement to the entries in the first row:

$$18 + 2 = 20 \qquad 23 + 2 = 25 \qquad 25 + 2 = 27 \qquad 27 + 2 = 29.$$

To obtain any remaining rows, successive y-decrements are added to the entries in the previous row:

$$20 + 1 = 21 \qquad 25 + 1 = 26 \qquad 27 + 1 = 28 \qquad 29 + 1 = 30.$$

As a check on our computations, we subtract the smallest y-observation from the largest x-observation. The result should be equal to the last entry in the last row. Indeed, we have $95 - 65 = 30$.

112 *Graphical Determination*

Unless the sample sizes m and n are quite small, the following graphical procedure for ordering the differences $x_i - y_j$ is usually faster than the arithmetic scheme in Section 111. This is particularly true when we are interested in finding a confidence interval. On a sheet of graph paper, we mark $x_1, \ldots, x_m$ along the horizontal axis and $y_1, \ldots, y_n$ along the vertical axis. We then plot the mn points with coordinates (x_i, y_j). Ordering the differences $x_i - y_j$ according to increasing size is equivalent to ordering the points (x_i, y_j) by sliding a 45°-line from the upper left to the lower right. The kth point in this ordering gives the kth smallest difference $x_i - y_j$.

The necessary steps for the graphic solution of the earlier example are indicated in Figure 14.13. Sliding a 45°-line from the upper left, we find that the 20th point has coordinates (70, 73) or (72, 75), both of which give $x - y = -3$. There are four choices for the 20th point when sliding a 45°-line from the lower right. We can choose any one of these four points, for example, the point (95, 75) which gives the value $x - y = 95 - 75 = 20$.

113 *Point Estimates of* Δ

We have observed that every difference $x_i - y_j$ furnishes a point estimate of Δ. The most appropriate of all these mn estimates is the one in the middle, that is, the median of the mn differences $x_i - y_j$.

FIGURE 14.13 Graphical Determination of Shift Parameter Estimates

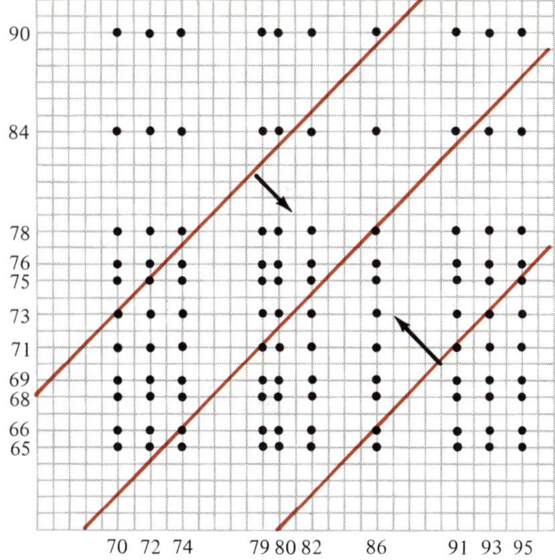

For the data in Table 14.1, there are 110 such differences. Our point estimate of Δ then lies halfway between the 55th and 56th difference. From Table 14.12 we find that both the 55th and 56th difference equal 8. Similarly, both the 55th and 56th point in Figure 14.13 fall on the same line. One of the points has coordinates (79, 71), corresponding to the difference $79 - 71 = 8$. Using either method, we find that the point estimate of Δ is 8.

PROBLEMS

1 Most Egyptian pyramids contain treasure chambers, practically all of which were discovered and plundered centuries ago. But no such treasure chamber has ever been found in the huge Chephren Pyramid. Recently scientists hypothesized that the possible existence of such a chamber in the Chephren Pyramid could be established by measuring the rate at which certain minute particles from outer space arrive in a vault under the pyramid. If there is an undiscovered chamber, the arrival rate of these particles would be higher than if the particles had to travel through solid rock. It is then necessary to compare measurements of arrival rates taken in a vault underneath the Chephren Pyramid with measurements of arrival rates of particles that are known to have traveled through solid rock.

a. Formulate the null hypothesis to be tested.
b. Specify the alternative hypothesis.

2 Given the following sets of observations:

x	14.2	12.7	9.6	13.1	
y	8.6	12.2	11.2	10.1	9.2

Compute

a. U_x;
b. U_y;
c. R_x;
d. R_y.

3 Check the value of U_y in Example 14.7 by computing R_y. Which method of computing U_y seems more convenient?

4 Find R_x and R_y for the data in Table 14.1. Check the values of U_x and U_y.

5 Use the data in Problem 21 of Chapter 12 to test the hypothesis that male and female drivers obtain the same mileage from a set of tires.

6 A testing company compared gasoline mileage (miles per gallon) of two cars of the same make in city driving. One car was equipped with manual shift, the other with automatic shift. Six trials for each car produced the following figures,

Manual Shift	18.2	19.5	17.9	18.0	18.7	19.1
Automatic Shift	17.4	16.8	17.5	17.0	16.2	16.5

Find a confidence interval for the difference in gasoline mileage.

7 A scientist is interested in comparing the body weight of two strains of laboratory mice. He observes the following weights (in grams),

Strain 1	37	32	43	35	41	37	31	35	31	37	36	29
Strain 2	37	43	33	45	47	51	37	43	41			

Do the two strains differ in body weight?

8 An experiment is run to determine which of two diets causes a greater weight gain in test animals. A group of 20 test animals is divided at random into two groups, the first receiving diet *A*, the second diet *B*. The amounts gained (in pounds) in a certain time interval under the two diets are,

Diet A	−1.0	0.0	2.1	3.1	3.3	4.3
	5.2	5.5	6.7	6.8		
Diet B	2.0	3.0	4.0	5.7	6.0	6.9
	7.0	7.2	7.3	8.1		

Do the two diets produce different weight gains?

9 The following are the grade-point averages of twenty high school seniors, ten of whom had the free use of a car during their senior year while the other ten did not have such privileges,

With Car	1.9	1.5	3.3	2.3	2.9	2.8	2.2	2.1	1.7	1.1
Without Car	3.2	3.5	3.9	3.8	2.3	3.4	1.9	3.0	1.8	2.5

Do these data bear out the claim that students who have free use of a car establish worse academic records than students without a car?

10 An automobile insurance company experiences the following claims (in dollars) for each of eight cars of two makes,

Make A	353	597	634	696	813	649	593	658
Make B	453	527	568	228	725	523	568	155

a. Do differential premium rates seem justified for the two makes?
b. Find a point estimate for the median difference in claims.

11 An automobile insurance company classified its policyholders as smokers and nonsmokers. For eight groups of 100 smokers each, the accident frequency in one year were as follows:

7 9 5 13 8 11 8 6

The corresponding rates for twelve groups of nonsmokers were:

4 2 4 3 3 6 3 4 2 4 3 2

Do these data support the claim that nonsmokers have significantly fewer accidents than smokers?

12 The grade-point averages of 12 male and 12 female seniors at a certain college are as follows,

Male	3.72	2.69	1.63	2.64	0.96	2.98
	1.95	2.24	3.71	2.61	1.46	3.27
Female	3.13	3.29	3.51	3.48	3.70	3.42
	3.25	1.92	3.64	2.87	2.54	1.90

Test the hypothesis that grade-point averages for male and female seniors do not differ. Use the

a. Median test.
b. Wilcoxon two-sample test.
c. What is the descriptive level associated with the Wilcoxon statistic?

13 On a certain Sunday, seven American League baseball games ended in the following scores:

5–4 6–4 6–2 5–3 10–7 6–3 8–6

while seven National League games ended in the following scores:

7–3 8–2 2–1 6–2 9–1 4–0 2–0

a. Test the hypothesis that the winning margin is the same in both leagues.
b. Consider all 14 scores as a random sample from the population of league baseball scores. Find a confidence interval ($\gamma = .99$) for the median winning margin in league baseball.
c. Test the hypothesis that winning scores in both leagues are the same.
d. Test the hypothesis that losing scores in both leagues are the same.

14 Consider the data in Table 15.1 giving the grades of three groups of pupils. Use the data for groups 2 and 3 only, and

a. test the hypothesis that the two methods of instruction do not produce different results;
b. find a confidence interval for a possible difference in test results.

15 A random sample of nine families in community A reported the following yearly incomes: 9100, 10300, 6800, 13500, 5500, 7900, 11000, 14000, 7200. A random sample of eight families in community B reported the following yearly incomes: 11500, 8300, 6100, 19000, 14500, 7200, 10500, 9400.

a. Would you say that the two communities differ with respect to family income?
b. Find a confidence interval for the difference in family income for the two communities.
c. In what sense do the answers under parts a and b agree?

16 It is claimed that the nicotine content of cigarettes of brand Y is lower than that of cigarettes of brand X. A laboratory makes the following determinations of nicotine content (in milligrams):

Brand X	1.0	1.3	1.5	1.1	1.6
Brand Y	0.8	1.2	1.4	0.9	1.0

Do you agree with the claim? Why?

17 On eight occasions of cloud seeding, the following amounts of rainfall were observed: .74, .54, 1.25, .27, .76, 1.01, .49, .70. On six control occasions (when no cloud seeding took place), the following amounts of rainfall were measured: .25, .36, .42, .16, .59, .66. Would you feel justified in claiming that cloud seeding increases amount of rainfall?

18 On the basis of 20 x-observations and 24 y-observations, you have found that $U_x = 188$. What is the descriptive level associated with this result against a

a. one-sided alternative;
b. two-sided alternative

19* Prove that $R_x + R_y = (m + n)(m + n + 1)/2$ even when ties are present.

20* Prove that $R_x = U_x + m(m + 1)/2$ and $R_y = U_y + n(n + 1)/2$. (*Hint:* Assume first that all x-observations are smaller than all y-observations.)

21* Find the distribution of U for $m = n = 3$ and check the entries in Table G.

15

COMPARATIVE EXPERIMENTS: *k* SAMPLES

114 In preceding chapters we considered the problem of how to compare two different methods or treatments and how to decide which, if either, is better. Often more than two methods are available. We want to generalize our test procedures to cover the comparison of any number of methods. For example, in evaluating the effectiveness of programmed instruction in high school algebra, we may give the same examination to three different groups of students:

Group 1: Students who have had programmed instruction but no supplementary discussions.
Group 2: Students who have had programmed instruction and, in addition, have had the opportunity to discuss the material with a qualified instructor.
Group 3: Students who have had regular classroom instruction but have not used programmed materials.

By considering, in addition, instruction by television, we could easily enlarge the scope of the experiment, ending up with considerably more than three groups of students.

The Kruskal-Wallis Test

115 Let us look at a set of examination scores involving three groups of students. We will assume that there are seven pupils in each group. As we will see later, our method of analysis can be used for any number of groups and any number of observations in each group. In an evaluation of various methods of teaching high school algebra, we may start with an equal number of pupils in each group, realizing that in the

course of the school year some pupils may drop out. It would be a great waste of effort, time, and money if the results for the remaining pupils could not be used to compare the various teaching methods.

Our basic data are presented in Table 15.1. As in the alternate approach to the Wilcoxon two-sample test, we replace observations by ranks—that is, we replace the smallest among the 21 observations by 1, the next smallest by 2, and so on, using midranks where ties occur.

TABLE 15.1 **Grades for Three Groups of Students**

Group 1	*Group 2*	*Group 3*
66	77	79
68	68	83
80	67	78
72	85	92
74	89	98
57	74	69
75	61	91

The results of the rankings, together with the sum of ranks for each group, are given in Table 15.2. The hypothesis that the teaching methods do not differ (as measured by the results of the examination) is rejected if the sums of the ranks for the three groups differ sufficiently from each other.

TABLE 15.2

Group 1	*Group 2*	*Group 3*
3	12	14
5.5	5.5	16
15	4	13
8	17	20
9.5	18	21
1	9.5	7
11	2	19
53	68	110

For the Wilcoxon test, it is possible to look in an appropriate table and decide whether the observed rank sums differ sufficiently from one another to suggest rejection or acceptance of the hypothesis being tested. For more than two rank sums corresponding tables are very awkward, so we rely on something along the lines of a chi-square test. In our example the ranks range from 1 to 21, so that the average rank is 11. Since each rank sum contains seven terms, if there is no difference among teaching methods each rank sum has expectation 77. A suitable quantity for measuring deviation from expectation is given by the sum of squares

$$(53 - 77)^2 + (68 - 77)^2 + (110 - 77)^2 = 1746.$$

While theoretically it would be possible to have a table from which we could find out whether to accept or reject our hypothesis, we can make use of the standard chi-square distribution with two degrees of freedom simply by multiplying the preceding sum of squares by a suitable constant. We will see that the hypothesis that the three teaching methods produce identical results can be rejected at the .05 level. It would appear that method 3 produces higher test scores than either method 1 or 2.

116 We now formulate the problem and its solution in general terms. First we define our notation. The number of treatments to be compared is k, where k is a number greater than 2. There are n_i observations for the ith method, $i = 1, \ldots, k$. The total number of observations is $N = n_1 + \cdots + n_k$. The sum of the ranks for the ith method is denoted by R_i. The test of the null hypothesis that there exist no differences among the k methods is based on the statistic

$$H = \frac{12}{N(N+1)}\left[\frac{1}{n_1}\left(R_1 - n_1\frac{N+1}{2}\right)^2 + \cdots + \frac{1}{n_k}\left(R_k - n_k\frac{N+1}{2}\right)^2\right],$$

which is more easily computed in the form

$$H = \frac{12}{N(N+1)}\left[\frac{R_1^2}{n_1} + \cdots + \frac{R_k^2}{n_k}\right] - 3(N+1).$$

The hypothesis being tested is rejected if the value of H is sufficiently large according to the chi-square distribution with $k - 1$ degrees of freedom. This test is known as the Kruskal-Wallis test.

The formula for H is particularly simple when $n_1 = \cdots = n_k = n$. Then

$$H = \frac{12}{nN(N+1)}[R_1^2 + \cdots + R_k^2] - 3(N+1).$$

For the data in Table 15.1, we find (using the rank sums in Table 15.2) that

$$H = \frac{12}{7 \times 21 \times 22}(53^2 + 68^2 + 110^2) - 3 \times 22 = 6.5.$$

For two degrees of freedom and significance level .05, Table D gives the critical value as 5.99. Since the observed value of H surpasses this critical value, the null hypothesis that the three teaching methods are equally effective should be rejected, as we indicated earlier.

Let us look at one further application of the Kruskal-Wallis test. A testing company compared five brands of automobile tires with respect to the distance required to bring a car traveling at a given speed on wet pavement to a full stop after locking the brakes. Table 15.3 gives the observed stopping distances in feet.

TABLE 15.3 **Braking Distances for Five Brands of Tires**

		Brand of Tire		
A	*B*	*C*	*D*	*E*
151	157	135	147	146
143	158	146	174	171
159	150	142	179	167
152	142	129	163	145
156	140	139	148	147
			165	166

To test the hypothesis that no differences exist, we convert the $N = 27$ observations to ranks and sum the ranks for each brand. This is done in Table 15.4. The computation of the test statistic H is now straightforward:

$$H = \frac{12}{27 \times 28}\left(\frac{75^2}{5} + \frac{60.5^2}{5} + \frac{21^2}{5} + \frac{120.5^2}{6} + \frac{101^2}{6}\right) - 3 \times 28 = 12.3.$$

This value of H surpasses the .025 critical value of chi-square with four degrees of freedom. Thus we reject the null hypothesis. The braking distance is not the same for the different brands of tires.

TABLE 15.4

		Brand of Tire		
A	*B*	*C*	*D*	*E*
15	18	2	11.5	9.5
7	19	9.5	26	25
20	14	5.5	27	24
16	5.5	1	21	8
17	4	3	13	11.5
			22	23
75	60.5	21	120.5	101

In practice, we often want more precise information. The H-test simply tells us that there are differences among the various types of tires. A tire user would want to know how various brands differ in braking ability. The original data indicate that any differences will not be large. But when it comes to braking ability, even a few feet may spell the difference between an accident and a safe ride. There is not much doubt that brand C is better than brand D. What about brands B and C, or A and C? Suppose that brand A is considerably cheaper than brand C. Should a car owner in need of new tires buy brand A or brand C tires (assuming that braking ability is the deciding factor)? The method of multiple comparisons provides answers to this and similar questions.

117 *Multiple Comparisons*

When comparing several treatments, we can decide which treatments differ from which (if any) by application of the method of multiple comparisons. For $i = 1, \ldots, k$, we denote the sum of the ranks corresponding to the ith treatment by R_i and let $r_i = R_i/n_i$ denote the average rank. Let $j = 1, \ldots, k$ and $i \neq j$. Intuitively, we would say that treatments i and j do not differ if $r_i - r_j$ is sufficiently close to zero. On the other hand, if $r_i - r_j$ takes a sufficiently large positive value, we would say that treatment i produces larger measurements than treatment j with a corresponding statement for large negative differences.

To make these statements more precise, we choose a significance level α such that we are willing to tolerate probability at most α of declaring that two or more of the k treatments differ, when in fact all k treatments are identical. When we use the Kruskal-Wallis test, we carry out just one overall comparison of all k treatments. It is then appropriate in most cases to use a rather small significance level like .05 or even .01. However, in the present approach we are interested in a large number of paired comparisons. Indeed, if we compare every treatment with every other treatment, the total number of possible comparisons is $k(k - 1)/2$, a number tabulated in Table 15.5 for k from 2 to 10. It is then reasonable to tolerate a considerably larger probability of making at least one false decision that two given treatments differ when in fact they do not. Thus we may very well choose α as large as .20 or even .25.

TABLE 15.5 Number of Paired Comparisons

k	2	3	4	5	6	7	8	9	10
$k(k-1)/2$	1	3	6	10	15	21	28	36	45

For a given value α, we can achieve the aim of having probability at most α of declaring that two or more treatments differ, when in fact all k treatments are identical in the following way. In Table C (or by interpolation in Table B) we find the value z that corresponds to the upper tail probability $\alpha' = \alpha/k(k - 1)$. For every pair i, j with $i \neq j$, we compute

$$z_{ij} = \frac{r_i - r_j}{\sigma_{ij}}$$

where

$$\sigma_{ij} = \sqrt{\frac{N(N+1)}{12}\left(\frac{1}{n_i} + \frac{1}{n_j}\right)}$$

$$= \sqrt{\frac{k(N+1)}{6}}, \qquad \text{if } n_1 = \cdots = n_k.$$

Finally, we declare that treatment i tends to produce smaller measurements than treatment j if $z_{ij} < -z$; larger measurements if $z_{ij} > z$; and similar measurements if $-z \leq z_{ij} \leq z$.

Let us return now to the tire problem where we wanted to compare five brands of tires pair by pair with respect to braking ability. Suppose that we are willing to tolerate an α-probability of .25. Then $\alpha/k(k-1) = .25/20 = .0125$ and $z = 2.241$. We first compute the rank averages $r_i = R_i/n_i$, where the rank sums R_i are taken from Table 15.4:

Brand	A	B	C	D	E
Rank Sum	75	60.5	21	120.5	101
Sample Size	5	5	5	6	6
Rank Average	15.0	12.1	4.2	20.1	16.8

Our earlier instructions were to compare ratios $(r_i - r_j)/\sigma_{ij}$ with the critical z-value, 2.241. An equivalent, and in the present case more convenient, procedure is to compare differences $r_i - r_j$ with the product $z\sigma_{ij}$. There are three different σ_{ij}'s depending on the sample sizes involved. Thus we find:

Sample Sizes	σ_{ij}	$z\sigma_{ij}$
5 and 5	5.02	11.25
5 and 6	4.81	10.78
6 and 6	4.58	10.26

We can now say that two brands differ in their performance if the difference in rank averages surpasses the appropriate value for $z\sigma_{ij}$. It follows that brand E can be expected to require greater stopping distances than brand C since the rank difference, $16.8 - 4.2 = 12.6$, surpasses the appropriate factor 10.78. Similarly, brand D can be expected to require a greater stopping distance than brand C. However, these are the only two pairs that differ significantly in their performance. Every other difference $r_i - r_j$ is numerically smaller than the appropriate factor $z\sigma_{ij}$. In making these statements, we should remember that we are running a risk of .25 of declaring at least one significant difference when in fact there are none.

118*

The determination of the appropriate z-value to carry out the multiple comparison procedure discussed in Section 117 is not as mysterious as it may appear to be at first. Our intention is to have overall probability at most α of declaring two or more treatments significantly different when in fact all k treatments have identical effects. Suppose that we carry out all $k(k-1)/2$ paired comparisons using a two-sided significance level α'' for each individual comparison. Thus α'' is the probability that the comparison between two specific treatments will be declared significant even though in reality the two treatments have identical effects. According to the generalization of Formula (3.9), the probability that at least one of the $k(k - 1)/2$ treatment pairs will be declared significantly different is at most equal to the sum of the individual probabilities. Thus if the null hypothesis is true, we have

$$P(\text{at least one pair of treatments is declared significantly different})$$

$$\leq \alpha'' + \cdot \cdot \cdot + \alpha'' = \frac{1}{2}k(k - 1)\alpha'' = k(k - 1)\alpha'$$

where $\alpha' = \frac{1}{2}\alpha''$. The probability on the left is $\leq \alpha$ if $k(k - 1)\alpha' = \alpha$. By choosing $\alpha' = \alpha/k(k - 1)$, we make sure that the overall probability of at least one erroneous decision does not surpass α when the null hypothesis is true.

The Friedman Test

119 In experiments we often obtain data that look as if they could be analyzed by means of the Kruskal-Wallis test. In Table 15.6 we have the average number of miles per gallon of gasoline that cars built by three different manufacturers achieved in a well-publicized economy run. We should like to have an answer to the following question: Do the three manufacturers achieve any consistent differences in gasoline mileage irrespective of car model? It may appear that we have the same kind of data as in Table 15.1. Again there are three treatments—the three manufacturers of cars—but outward appearance is deceptive. It would be a serious mistake to analyze the data in Table 15.6 by the H-test. Let us see why.

TABLE 15.6 Gasoline Mileage for Various Cars

	Manufacturer		
Model	G	F	C
Compacts	20.3	25.6	24.0
Intermediate 6s	21.2	24.7	23.1
Intermediate 8s	18.2	19.3	20.6
Full-size 8s	18.6	19.3	19.8
Sports Cars	18.5	20.7	21.4

In the present case our observations appear in distinct groups of three, one group for each model class. This is quite different from the setup of Example 15.1, where pupils were not matched in groups. What we are considering are generalizations of the two different two-sample setups discussed in Chapters 13 and 14. The Kruskal-Wallis test generalizes the Wilcoxon test of Chapter 14 for independent samples.

What we need now is a test procedure that is appropriate when observations are collected in groups of k, each group containing exactly one observation for each of the k treatments under investigation. This is a generalization of the setup of Chapter 13, when observations occurred in pairs. Since observations belonging to the same group are generally related, the assumption of independent samples underlying the Kruskal-Wallis test is likely to be violated.

For the Kruskal-Wallis test, we assign ranks within the set of all observations to compare every observation with every other observation. Under the present setup, it is necessary to assign ranks separately within each group. (It certainly makes little sense to compare gasoline mileage of compacts with gasoline mileage of full size V8s.) We then find the following rankings:

	G	F	C
Compacts	1	3	2
Intermediate 6s	1	3	2
Intermediate 8s	1	2	3
Full-size 8s	1	2	3
Sports Cars	1	2	3
	$R_1 = 5$	$R_2 = 12$	$R_3 = 13$

As in the case of the Kruskal-Wallis test, we compute the sum of ranks for each treatment. However, since we have used a different ranking procedure, these rank sums have to be analyzed differently. For k treatments and n groups, the appropriate test statistic is

$$Q = \frac{12}{nk(k+1)}\left\{\left[R_1 - \frac{1}{2}n(k+1)\right]^2 + \cdot\ \cdot\ \cdot + \left[R_k - \frac{1}{2}n(k+1)\right]^2\right\}$$

$$= \frac{12}{nk(k+1)}[R_1^2 + \cdot\ \cdot\ \cdot + R_k^2] - 3n(k+1).$$

The hypothesis that there are no differences among the k treatments is rejected if the value of Q surpasses the tabulated value of chi-square with $k - 1$ degrees of freedom at the chosen significance level.

For the gasoline mileage data, $k = 3$ and $n = 5$, so that

$$Q = \frac{12}{5 \times 3 \times 4}(5^2 + 12^2 + 13^2) - 3 \times 5 \times 4 = 7.6,$$

a value that indicates rejection of the null hypothesis at the .02 level. It would seem that the cars produced by manufacturer G deliver lower gasoline mileage than the cars produced by manufacturers F and C.

Our intuitive judgment is borne out by the method of multiple comparisons. For the present setup, we compare rank sum differences $R_i - R_j$ with $z\sigma$, where $\sigma = \sqrt{nk(k+1)/6}$.

Suppose that we choose $\alpha = .10$ for our gasoline mileage data. Then we find that $\alpha/k(k-1) = .10/6 = .0167$ and therefore $z = 2.13$. Further, we have $\sigma = \sqrt{5 \times 3 \times 4/6} = 3.16$, so that $z\sigma = 6.73$. Since both $R_2 - R_1$ and $R_3 - R_1$ are greater than 6.73, we conclude that the gasoline mileage of cars manufactured by G is lower than that of cars manufactured by either F or C.

Completely Randomized and Randomized Block Designs

120 We have discussed two different tests involving the comparison of k treatments. The two tests are not interchangeable. Which of the two tests is applicable in a given situation, if either, depends on the way the experiment that produces the basic data is designed. Suppose that we have N experimental units available for our experiment. The 21 pupils in the teaching experiment are the experimental units. The 15 cars in the gasoline mileage experiment are the experimental units. The Kruskal-Wallis test is appropriate whenever the N available experimental units are assigned completely randomly to the k treatments, n_1 to treatment 1, n_2 to treatment 2, . . . , n_k to treatment k. We then have a *completely randomized design.*

On the other hand, it often happens in practice that the available $N = kn$ experimental units are, or can be, divided into n homogeneous groups, or *blocks* as we will say from now on, each block consisting of k units. In such a case, all k treatments should be used exactly once (in random order) in each of the n blocks, resulting in a so-called *randomized block design.* Such data are analyzed by the Friedman procedure.

The second experimental setup calls for some further comments. In many experiments blocks arise in a natural way. Thus in the gasoline mileage example the division of experimental units into blocks is imposed by the various model classes. But in other experiments the experimenter may have to make a conscious effort to form appropriate blocks (see Problem 9 of this chapter). The problem of whether to form blocks or not is discussed in detail in courses on the design of experiments. Here we limit ourselves to one or two comments. Clearly, a block design of the type we have discussed requires that the number of experimental units be a multiple of the number of treatments. However, this does not mean that whenever this condition is satisfied, we should try to arrange for blocks. It is advisable to use a block design only if, within blocks, greater homogeneity of the factors having a bearing on the phenomenon being investigated can be achieved.

121 *The Problem of n Rankings*

The Friedman test is often used to solve a problem known as the *problem of n rankings.*† In this problem n persons, often referred to as *judges,* are asked to rank k objects (like contestants for a prize, brands of consumer goods, and so on) in order of preference. Here the judges correspond to blocks; the objects, to treatments. The purpose of such an experiment is to find out if there is some agreement among the n judges with respect to their order of preference. (In this sense the problem of n rankings has some similarity with one of the topics Chapter 16 discusses.) Agreement among the judges is indicated by a high value of Q. On the other hand, a low value of Q may mean one of two things: pronounced disagreement among the judges, or random arrangement of the objects due to a lack of preference. This second possibility corresponds to the earlier null hypothesis that k treatments do not produce different effects. If the objects to be ranked by the judges differ little or not at all, the judges are likely to arrange them in random order.

Paired Comparisons

122 When the number k of treatments that we want to compare is large, the only kind of homogeneous blocks that can be found may contain fewer than k units. This has lead to the development of *incomplete block designs.* We consider only one simple design and its analysis.

We have mentioned consumer preference ratings as an application of the problem of n rankings. Experience shows that consumers provide the most reliable information if each consumer is asked to compare only two brands. A consumer participating in such an experiment represents one block with two experimental units. The entries in Table 15.5 can then also be interpreted as the number of consumers needed to obtain one complete set of comparisons.

EXAMPLE 15.7 Four brands can be compared using six consumers as follows:

Consumer	*A*	*B*	*C*	*D*
1	∗	∗		
2	∗		∗	
3	∗			∗
4		∗	∗	
5		∗		∗
6			∗	∗

Here an ∗ indicates that the consumer will be asked to rate the indicated brand and compare it with the other starred brand. If $6r$ consumers are available, the same setup is repeated r times.

†Many statisticians use the letter m in place of n when speaking of this problem. However, the letter n is more in line with our regular notation.

The analysis of data from such a paired comparison experiment is not complicated. In each row we mark 1 for the preferred brand and leave the other positions blank. This procedure corresponds to using "ranks" 0 and 1, rather than 1 and 2, when ranking the brands in each block. The effect is the same. Let R_i be the sum of the entries in the ith column. For a paired comparison experiment, R_i is the number of times brand i has been preferred in the various comparisons in which it has been used. To test the null hypothesis that consumers taken as a whole do not prefer any one brand over any other, we compute

$$Q = \frac{4}{rk}\left(\left[R_1 - \frac{1}{2}r(k-1)\right]^2 + \cdot \cdot \cdot + \left[R_k - \frac{1}{2}r(k-1)\right]^2\right)$$

$$= \frac{4}{rk}[R_1{}^2 + \cdot \cdot \cdot + R_k{}^2] - r(k-1)^2,$$

where r equals the number of complete comparison sets, a complete set requiring $k(k-1)/2$ consumers. The hypothesis is rejected if the observed value of Q surpasses the tabulated critical value of chi-square with $k - 1$ degrees of freedom.

Multiple comparisons are performed by comparing rank sum differences $R_i - R_j$ with $z\sigma$ where $\sigma = \sqrt{rk/2}$.

EXAMPLE 15.8 Forty consumers participated in a paired comparison experiment involving the comparison of five brands of instant coffee. Here are their preferences:

$$R_1 = 7, \qquad R_2 = 14, \qquad R_3 = 11, \qquad R_4 = 3, \qquad R_5 = 5.$$

Thus brand 1 was preferred by 7 of the consumers who compared brand 1 with some other brand, brand 2 was preferred by 14 consumers, and so on. Can we conclude from these data that consumers have definite preferences among the various brands of instant coffee or are they simply making random choices?

Since we are comparing five brands, we have $k = 5$. According to Table 15.5, a complete set of comparisons requires 10 consumers. Since 40 consumers participated in the experiment, we have $r = 4$. Then

$$Q = \frac{4}{4 \times 5}(7^2 + 14^2 + 11^2 + 3^2 + 5^2) - 4 \times 4^2 = 16.$$

With 4 degrees of freedom, the result is highly significant. Apparently consumers are not making random guesses. We may then ask which of the brands are preferred to which. The multiple comparison method provides an answer. Using $\alpha = .20$ for a change, we find that $\alpha/k(k-1) = .20/20 = .01$, and therefore $z = 2.33$. Since σ equals $\sqrt{4 \times 5/2} = 3.16$, we have $z\sigma = 7.36$. Looking at differences $R_i - R_j$, we conclude that consumers prefer brand 2 to brands 4 and 5, and brand 3 to brand 4. However, when it comes to other pairs of brands, consumers do not appear to have any common preferences.

PROBLEMS

1 Suppose that you want to compare the performance of six different suntan lotions using the services of the eight test subjects in Chapter 13. Design an appropriate experiment and indicate how you would analyze the resulting data.

2 Suppose that in Table 15.3 brands A and C are radial tires while brands B, D, and E are conventional tires. Use the appropriate two-sample procedure to test the hypothesis that there is no difference in the stopping distances of the two types of tires.

3 Compute the statistic H for the following data:

Sample 1	40	45	30	42	38
Sample 2	39	33	44		
Sample 3	47	43	49	55	

4 Compute the statistic Q for the following data:

	Treatments			
	19	14	11	17
Blocks	16	15	12	18
	19	11	12	16

5 Twenty test subjects each using one of four reducing diets lost the following number of pounds over a period of four weeks:

Diets

A	*B*	*C*	*D*
2	9	4	18
7	10	17	14
22	8	0	24
18	20	6	13
15		11	12
			22

Is there any difference in the effectiveness of the four diets to produce loss of weight?

6 Six movie critics were asked to arrange four pictures A, B, C, and D in order of preference. The results were as follows:

Critic

1	2	3	4	5	6
B	*B*	*A*	*B*	*A*	*B*
A	*C*	*B*	*C*	*B*	*C*
C	*A*	*C*	*D*	*D*	*A*
D	*D*	*D*	*A*	*C*	*D*

(For example, critic 1 preferred picture B over A over C over D.) Set up an appropriate hypothesis and test it.

7 In a paired comparison experiment, 30 consumers rated four kinds of soft drinks as follows: 12 preferred brand 1; 7, brand 2; 3, brand 3; and 8, brand 4. Is there any kind of consensus among the consumers?

8 Plan and carry out a paired comparison experiment. Analyze your data as completely as possible.

9 The experiment for comparing three methods of teaching algebra (Table 15.1) could have been designed differently. If sufficient information about the pupils is available, we can divide 21 pupils into seven groups of three in such a way that the three students in each group are as similar as possible with respect to factors thought to have a bearing on a pupil's learning ability for algebra. In each group one student is randomly assigned to method 1, another to method 2, and a third to method 3. Suppose that the examination scores for this experiment are as follows:

Group	*Method 1*	*Method 2*	*Method 3*
1	69	68	79
2	75	85	92
3	57	61	78
4	72	78	91
5	80	89	98
6	66	70	69
7	74	73	83

What are your conclusions?

10 Twenty-one students took advanced GRE (Graduated Record Examination) tests, seven each in history, mathematics, and sociology. They obtained the following scores:

History	550	540	490	620	610	550	570
Mathematics	820	760	670	500	690	540	910
Sociology	510	550	570	380	420	640	490

Would you say that the achievement level of students taking the three types of examinations is the same?

11 In a supermarket consumer survey, customers were asked to rank the following complaints according to their importance to the customer:

A Leaky milk cartons
B Packages that do not tear where indicated
C Spray cans that do not spray
D Unintelligible directions for use
E "Economy" size packages that are not economical
F Advertized specials that are unavailable

Four customers arranged their complaints as follows (from most important to least important):

Customer	1	*C*	*A*	*F*	*B*	*D*	*E*
	2	*A*	*F*	*B*	*C*	*E*	*D*
	3	*D*	*C*	*B*	*F*	*E*	*A*
	4	*B*	*F*	*D*	*C*	*A*	*E*

Would you say that the four customers tend to agree as to the importance of the various complaints?

12 A testing organization rated three brands of tires on each of five characteristics on a scale from 1 (low performance) to 10 (high performance):

	Brand		
	A	*B*	*C*
Cornering	10	8	7
Wet skid	10	9	7
Tracking	6	10	8
Braking	8	7	6
Wear	8	9	10

Considering all five characteristics, might we conclude that any particular brand seems best?

13 An instructor taught three sections of elementary statistics, one at 8 o'clock, one at 10 o'clock, and a third at 1 o'clock. The students in all three classes took the same final examination. Their grades were:

8 o'clock	72	84	75	90	97	85	76	87		
10 o'clock	77	66	79	55	71	69	73	82		
	83	67	70							
1 o'clock	58	78	80	62	68	74	81	56	42	65

What conclusions do you draw from these data?

14 A consumer testing organization tested the performance of six electric shavers. The organization assembled a users' panel of sixty men and gave each member of the panel two shavers (using each possible combination of two shavers an equal number of times). The panel members were asked to use the shavers on alternate days for two weeks and then report which shaver they preferred. The results were:

Shaver	*A*	*B*	*C*	*D*	*E*	*F*
Number of Preferences	2	12	18	8	14	6

Do the panel members tend to agree in their preferences?

15 Thirty subjects in a consumer preference study were asked to arrange three pieces of pastry in order of preference. One piece had a light (L) crust, a second piece had a medium (M) crust, and the third piece had a heavy (H) crust. The following table gives, for each of the six possible orders, the number of subjects that chose the particular order (least preferred on the left, most preferred on the right):

Order	*Number of Preferences*
H M L	10
M H L	3
H L M	3
L H M	8
M L H	1
L M H	5

Is there agreement among the thirty subjects about the kind of pie crust they prefer?

16 A newspaper reporter took a car that had sustained considerable damage in a rear-end collision to 26 randomly selected garages in a given city and asked for repair estimates. Some garages were told that damage was covered by insurance, some garages were told that no insurance coverage was available, and the remaining garages were told nothing about possible insurance coverage. The repair estimates (in dollars) were as follows:

Insured	*Noninsured*	*No Insurance Information*
754	567	492
727	500	445
670	473	406
651	472	398
618	449	375
596	400	342
574	392	
559	345	
489	300	
469	212	

Set up an appropriate hypothesis and test it.

17 The Mobil Oil Corporation ran a newspaper ad to put recent increases in the price of gasoline "in perspective." The ad listed price changes between 1960 and 1973 of 14 items:

Item	*Average Price* 1960	1970	1974
Bacon (1 lb)	.64	.98	1.25
Bread, white (1 lb)	.20	.24	.34
Cigarettes (1 pkg)	.27	.41	.47
Classified ad (1 line) (*N.Y. Times*)	2.30	3.30	4.06
Dental care (1 filling)	5.08	7.33	8.99
Eggs (1 doz)	.56	.57	.78
Gasoline, regular (1 gal)	.31	.36	.54
Hose, women's (1 pair)	1.52	1.52	1.39
Movie admission (adult)	.95	1.81	2.16
Roast, rib (1 lb)	.81	1.12	1.55
Shoes, men's (1 pair)	15.24	20.40	24.71
Toilet tissue (650-sheet roll)	.09	.10	.13
Washing machine	239.11	226.83	238.19
Vitamins (100 capsules)	3.16	2.78	2.76

What statistical analysis does the table suggest?†

---

†Reprinted by permission of the Mobil Oil Corporation. © 1974 Mobil Oil Corporation.

18 A student organization surveyed food prices at four local food stores:

Item	*Weight/volume*	*Stores* A	B	C	D
Apples	per lb	.30	.30	.33	.45
Lettuce	one head	.39	.25	.25	.39
Milk, homogenized	½ gal container	.84	.76	.81	.76
Eggs: fresh, grade A, large	1 doz	.89	.83	.69	.93
Hamburger	per lb	1.29	.99	.99	1.09
Frying chicken	cut up, per lb	.65	.46	.59	.69
Chicken noodle soup	10¾ oz can	.22	.19	.22	.19
White bread	1 lb loaf	.48	.59	.48	.33
Raviolios with meat sauce	15 oz	.45	.41	.43	.35
Soda	qt bottle	.38	.40	.37	.39
Coffee	4 oz	1.39	1.31	1.29	1.23
Peanut butter	28 oz jar	1.19	1.16	1.17	1.09
Laundry soap	3 lb 1 oz	.89	.85	.81	.80

Suppose that you want to do all your shopping at one store. Does it make any difference at which of the stores you shop?

19 Apply the method of multiple comparisons to the data in Table 15.1.

20 Apply the method of multiple comparisons to the data in Table 15.3 using $\alpha = .20$.

21 Apply the method of multiple comparisons to problem 7 of this chapter.

22 Apply the method of multiple comparisons to problem 9 of this chapter.

23 Apply the method of multiple comparisons to problem 10 of this chapter.

24 Apply the method of multiple comparisons to problem 13 of this chapter.

25 Apply the method of multiple comparisons to problem 14 of this chapter.

26 Apply the method of multiple comparisons to problem 16 of this chapter.

27 Apply the method of multiple comparisons to problem 17 of this chapter.

28 Apply the method of multiple comparisons to problem 18 of this chapter.

29* Find the number of consumers needed to perform a paired comparison experiment involving

a. 11 treatments,
b. 12 treatments.

30* A chess tournament is planned in which each of ten players plays twice against every other player.

a. How many games does each player play?
b. How many games will be played in all?

16

RANDOMNESS, INDEPENDENCE, AND RANK CORRELATION

123 Implicit in all procedures discussed in earlier chapters is the assumption that observations in a sample are random. If observations are random, the order in which they are recorded is immaterial and may be ignored. However, there are situations when the order of successive observations is important. Suppose that we record the price of a certain commodity like the price of a gallon of gasoline at monthly intervals. In a stable economy successive observations will show random fluctuation, but the overall level will be fairly constant. If we live in an inflationary economy, there still will be random fluctuation from one month to another, but underlying this random fluctuation will be a gradual—or perhaps not so gradual—increase in price.

In Section 124 we will discuss a test that allows us to distinguish mere random fluctuation from random fluctuation superimposed on a monotone trend (upward or downward). The test is based on a statistic S that also plays a role in the discussion of independence and rank correlation. We describe the computation of S.

Let

$$y_1, y_2, \cdot \cdot \cdot, y_n$$

be a sequence of observations. With each such sequence we associate a statistic S computed as follows. We compare every observation in the sequence with every other observation further to the right. For any pair of observations y_i and y_j with $i < j$, we count $+1$ if $y_i < y_j$ and we count -1 if $y_i > y_j$. If $y_i = y_j$, we count 0 or, what amounts to the same thing, ignore the pair. For a sequence of n observations, there are

$$(n - 1) + (n - 2) + \cdot \cdot \cdot + 2 + 1 = \frac{1}{2}n(n - 1)$$

comparisons, since y_1 is compared with $n - 1$ observations further to the right; y_2 is compared with $n - 2$ observations further to the right; etc. The statistic S equals the sum of all $+1$s and -1s in the $\frac{1}{2}n(n-1)$ comparisons.

As an example, if the sample contains the four observations

$$y_1 = 4, \qquad y_2 = 6, \qquad y_3 = 1, \qquad y_4 = 4,$$

then we find that y_1 is smaller than y_2 for a count of $+1$, greater than y_3 for a count of -1, and equal to y_4 for a count of 0; y_2 is larger than both y_3 and y_4 for a combined count of -2; finally, y_3 is smaller than y_4 for a count of $+1$. Thus

$$S = (1 - 1) + (0 - 2) + (1 - 0) = -1,$$

where the three parentheses indicate how many $+1$s, and -1s are contributed by y_1, y_2, and y_3, respectively.

A Test of Randomness against a Monotone Trend

124 An upward trend in a sequence of observations $y_1, y_2, \ldots, y_n$ is characterized by a gradual increase in the size of observations as we go from the first observation on the left to the last observation on the right. Such a sample produces an excess of positive counts over negative counts; the result is a large positive value of S. A downward trend produces an excess of negative counts; the result is a large negative value of S. Finally, a random sequence has approximately equal numbers of positive and negative counts. The result is a value of S near zero. Large positive values of S suggest rejection of the hypothesis of randomness in favor of an upward trend, while large negative values of S suggest rejection of the hypothesis of randomness in favor of a downward trend. More specifically, the hypothesis that a sequence of observations is random against the alternative of a monotone trend is tested by computing

$$z = \frac{6S}{\sqrt{2n(n-1)(2n+5)}}$$

and comparing z with one- and two-sided critical values listed in Table C. We will often find it convenient to refer to this test as the S-test.

EXAMPLE 16.1 The order in which young men were to be inducted into the United States Armed Forces during 1970 was determined by a draft lottery in which numbers from 1 to 366 were assigned to individual birth dates. Men with draft numbers from 1–183 were much more likely to be called than men with draft numbers from 184–366. The following table gives the number of dates in the 1–183 category for each month of the year:

Jan.	Feb.	March	April	May	June
12	12	10	11	14	14

July	Aug.	Sept.	Oct.	Nov.	Dec.
14	19	17	13	21	26

In a properly conducted lottery, dates falling in the 1–183 category should be randomly distributed over the twelve months of the year. Let us apply our test of randomness to the given data. We find that

$$S = (8-2) + (8-2) + (9-0) + (8-0) + (4-1) + (4-1) + (4-1) + (2-2) + (2-1) + (2-0) + (1-0) = 42$$

and

$$z = \frac{6 \times 42}{\sqrt{2 \times 12 \times 11 \times 29}} = 2.88.$$

The descriptive level associated with $z = 2.88$ is .002 for one-sided alternatives and .004 for two-sided alternatives. There is overwhelming evidence of nonrandomness in the given data.

It is instructive to ask what went wrong with the 1970 draft lottery and we discover the answer by following the steps in the process. The 366 dates from January 1 through December 31 were written on separate slips of paper and each slip was put in a small capsule. After some mixing, the capsules were placed in a large glass bowl and were drawn out one at a time by various celebrities to determine the draft order of the 366 birth dates. As it turned out, the mixing of the capsules was quite insufficient to overcome the original ordering of the capsules. Any statistician familiar with random sampling procedures could have predicted what would happen, but none had been consulted. Actually, even though nonrandom, the draft lottery for 1970 provided an acceptable ordering of induction dates, since the birth date of an individual imparts the necessary element of randomness to the selection process. Plans for subsequent draft lotteries were prepared by professional statisticians.

We have pointed out before that ties among the observations cause theoretical complications. If nonparametric methods of analysis are used, it is good statistical practice to try to avoid ties among the observations as much as possible. Sometimes greater precision in making a measurement eliminates ties. However, in the draft example greater precision is impossible. The number of draft dates in any one month is necessarily an integer. Both January and February have exactly 12 draft dates, no more and no less. In such a case it may be possible to use additional criteria to impose an *objective* order on tied observations. In the draft example we note that January has 31 days while February has only 29 days. Thus 12 draft dates in February represent a higher concentration of draft dates than 12 draft dates in January. We may then change our original table to

Jan.	*Feb.*	*Mar.*	*Apr.*	*May*	*June*
12−	12+	10	11	14−	14+

July	*Aug.*	*Sept.*	*Oct.*	*Nov.*	*Dec.*
14−	19	17	13	21	26

where 12+ is considered to be larger than 12− and 14+ larger than 14−. While we have not been able to eliminate all ties, we are considerably better off than before. We now find that

$$S = (9 - 2) + (8 - 2) + (9 - 0) + (8 - 0) + (5 - 1) + (4 - 2) + (4 - 1) + (2 - 2) + (2 - 1) + (2 - 0) + (1 - 0) = 43$$

and

$$z = \frac{6 \times 43}{\sqrt{2 \times 12 \times 11 \times 29}} = 2.95,$$

an even more significant z-value than before.

A Test of Independence

125 The S-test of Section 124 can be used to solve another kind of problem. Suppose that we have the final examination grades of ten students in two subjects, psychology and statistics, as presented in Table 16.2. Such

TABLE 16.2

Psychology	92	65	82	76	84	77	69	75	97	72
Statistics	85	72	91	70	79	80	60	73	75	82

data can be represented graphically by a scatter diagram as in Figure 16.3. Inspection of Figure 16.3 suggests that, on the whole, students who have low grades in psychology also have low grades in statistics,

FIGURE 16.3

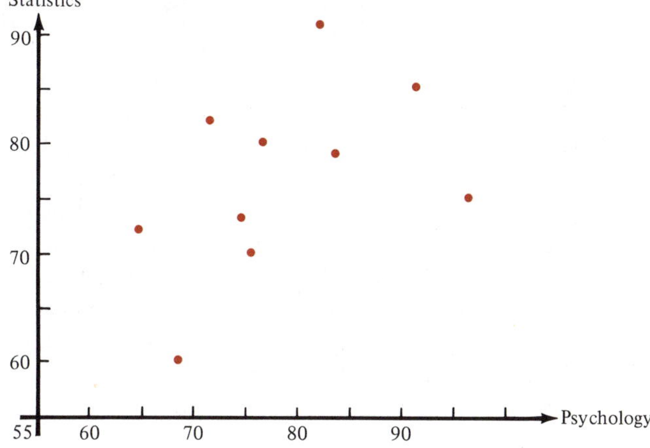

and students who have high grades in psychology also have high grades in statistics or vice versa. But there are exceptions like the student who has 97 in psychology and only 75 in statistics. We should like to know what we can say about students in general who take psychology and statistics. Are psychology grades and statistics grades related or unrelated? We will adapt the S-test of Section 124 to serve as a test of independence to answer this question.

In Section 124 the quantity S is used as an indicator of a possible trend among observations $y_1, \ldots, y_n$ relative to the order in which observations are taken (order indicated by the subscripts.) If we let $y_1, \ldots, y_{10}$ denote statistics grades in the present example, the subscripts 1 through 10 have no particular significance. They simply serve to identify ten students. But there is another set of measurements that can be used to impose an ordering on statistics grades, namely, psychology grades. The similarity will be clearer if we rearrange Table 16.2 in the form of Table 16.4. Now students have been ordered according to their grades on the psychology examination, and our earlier question concerning a possible relationship between psychology and statistics grades can be reformulated as to whether the rearranged statistics grades do or do not indicate an upward (or possibly, downward) trend to match the order of the psychology grades. This question can be answered by applying the S-test of Section 124 to the statistics grades as arranged in Table 16.4.

TABLE 16.4

Psychology	65	69	72	75	76	77	82	84	92	97
Statistics	72	60	82	73	70	80	91	79	85	75

We find that

$$S = (7 - 2) + (8 - 0) + (2 - 5) + (5 - 1) + (5 - 0) + (2 - 2) + (0 - 3) + (1 - 1) + (0 - 1) = 15$$

and $$z = \frac{6S}{\sqrt{2n(n-1)(2n+5)}} = \frac{6 \times 15}{\sqrt{2 \times 10 \times 9 \times 25}} = 1.34.$$

The descriptive level associated with this z-value (using a two-sided test) is $2 \times .090 = .180$, so that we cannot rule out the possibility that, for the student population as a whole, psychology and statistics grades are independent.

Let us now formulate this test of independence in general terms. We have observed n pairs of measurements (x, y) and should like to test the hypothesis that x- and y-measurements are unrelated against the alternative that x- and y-measurements tend to move in the same direction (as one increases, so does the other) or in opposite directions (as one increases, the other decreases).

To test for independence, we arrange the n pairs of observations according to the value of the variable x,

$$x_1 \leq x_2 \leq \cdot \quad \cdot \quad \cdot \leq x_n.$$

(Now subscripts refer to the size of the x-observations and not to the order in which the x-observations were obtained.) By y_1 we denote the y-observation that is paired with the smallest x-observation; by y_2, the y-observation that is paired with the second smallest x-observation; and so on. Next we compute the statistic S from the sequence

$$y_1 \quad y_2 \quad \cdot \quad \cdot \quad \cdot \quad y_n.$$

The hypothesis of independence is rejected provided that

$$z = \frac{6S}{\sqrt{2n(n-1)(2n+5)}}$$

falls in the critical region according to Table C.

EXAMPLE 16.5 To assess the "quality of life" in various United States cities, a research organization ranked cities with respect to a large number of factors. The following are the rankings of eight cities on housing and transportation (a ranking of 1 indicates most favorable, a ranking of 8, least favorable, conditions:

	Housing	*Transportation*
New York	7	1
Los Angeles	3	3
Chicago	5	6
Philadelphia	2	2
Detroit	1	4
San Francisco	6	8
Washington	4	5
Boston	8	7

We should like to know whether the two criteria, housing and transportation, are related.

We note that in the present example we do not have actual measurements of the quality of housing and transportation, only the ranking of the various cities relative to each other. One of the advantages of the statistical methods of this chapter is that they are applicable to rankings. The same is not true of the corresponding methods in Chapter 20.

If we use housing as the x-variable and transportation as the y-variable, we find the following rearranged data:

Housing	1	2	3	4	5	6	7	8
Transportation	4	2	3	5	6	8	1	7

We compute S from the rearranged transportation ranks:

$$S = (4 - 3) + (5 - 1) + (4 - 1) + (3 - 1) + (2 - 1) + (0 - 2) + (1 - 0) = 10.$$

Finally, we find that $z = 6 \times 10/\sqrt{16 \times 7 \times 21} = 1.24$, which is too small to reject the hypothesis of independence. (See, however, Problem 15.)

126* *Tied Observations*

In Section 124 when discussing the S-test of the present chapter as a test of randomness, we noted that pairs of tied y-values did not contribute anything when computing S. The same is true when the S-test is used as a test of independence, but now an additional problem may arise. There may be ties among the x-observations as well. Since x-observations are used to order y-observations, ties among x-observations introduce an element of indeterminancy when ordering y-observations. Thus for the following four pairs of observations (in each pair, the first number is the x-observation, the second number is the y-observation)

$$(2, 4) \qquad (2, 3) \qquad (5, 2) \qquad (1, 2),$$

there are two possible orderings of the y-observations:

Ordered x-observations	1	2	2	5	
Corresponding y-observations	2	4	3	2	or
	2	3	4	2.	

We may compute S for *each* of the possible y-arrangements and *average* the results. The first arrangement gives

$$S_1 = (2 - 0) + (0 - 2) + (0 - 1) = -1$$
$$S_2 = (2 - 0) + (1 - 1) + (0 - 1) = +1$$

and
$$S = \frac{1}{2}(S_1 + S_2) = 0.$$

A faster procedure is to select either y-arrangement and compute S as usual except that we ignore any pair of y-values corresponding to tied x-values. Thus whether we use the first or the second y-arrangement we find that

$$S = (2 - 0) + (0 - 1) + (0 - 1) = 0.$$

The pair (4, 3) in the first y-arrangement [(3, 4), in the second y-arrangement] is ignored when computing S, since it corresponds to tied x-values.

Rank Correlation

127 Past experience or theoretical considerations may suggest that two variables x and y are not unrelated. In such a case we are often interested in a measure of the strength of relationship between x and y. Statisticians have invented various *correlation coefficients* to measure strength of relationship. One of the most generally applicable and useful is Kendall's rank correlation coefficient.

The word *rank* in the name of the coefficient indicates that it can be computed from rankings as in Example 16.5. Indeed, our discussion is much simplified if we start with an example involving rankings. Suppose that two judges have ranked six contestants participating in a competition with the results as shown in Table 16.6. The two rankings

TABLE 16.6

Contestant	*A*	*B*	*C*	*D*	*E*	*F*
First Ranking	4	6	2	3	1	5
Second Ranking	3	6	1	4	2	5

show neither complete agreement nor complete disagreement. Complete agreement means identical rankings:

$$\begin{matrix} 1 & 2 & 3 & 4 & 5 & 6 \\ 1 & 2 & 3 & 4 & 5 & 6 \end{matrix}$$

Complete disagreement means reversed rankings:

$$\begin{matrix} 1 & 2 & 3 & 4 & 5 & 6 \\ 6 & 5 & 4 & 3 & 2 & 1 \end{matrix}$$

Most rankings fall between these two extremes. Our example represents one of the in-between situations. But looking at the two rankings, we have the impression that they tend more toward perfect agreement than toward complete disagreement. Can we support this subjective judgment by an objective, quantitative statement?

Again we compute S for the two rankings. We arrange contestants according to the ranking of the first judge:

Contestant	*E*	*C*	*D*	*A*	*F*	*B*
First Ranking	1	2	3	4	5	6
Second Ranking	2	1	4	3	5	6

and use the second ranking to compute S,

$$S = (4 - 1) + (4 - 0) + (2 - 1) + (2 - 0) + (1 - 0) = +11.$$

The result $S = +11$ characterizes the degree of relationship between the two rankings. How close is the result to perfect agreement or perfect disagreement? This question is answered by computing S for the rankings representing the two situations. In the first case, every pair contributes $+1$ for a total score of $+15$; in the second case, every pair contributes -1 for a total score of -15. Thus $S = +15$ represents perfect agreement; $S = -15$ represents complete disagreement.

We can exhibit our result graphically. On a scale extending from -15 to $+15$, our rankings correspond to a value of $+11$:

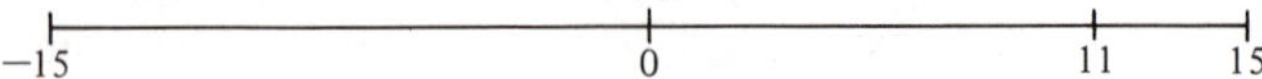

The example shows the disadvantage of S as a measure of strength of relationship. We gain an impression of the strength of relationship only by considering S relative to its two possible extremes. This deficiency is easily remedied. All we have to do is divide S by its maximum value, that is, consider the new quantity $t = S/(\max S)$ in place of S. Then t is measured on a scale going from -1 to $+1$. For our example, $t = {}^{11}\!/_{15} = .73$, a value which indicates clearly where t is located relative to the two extremes -1 and $+1$.

The quantity t is known as *Kendall's rank correlation coefficient.* A value of t near $+1$ implies close agreement among the rankings. A value of t near -1 implies almost diametrically opposite rankings. A value of t in the neighborhood of 0 indicates neither agreement nor disagreement. In this last case we may say that the two rankings are unrelated. Neither ranking can be used to throw light on the other. A more precise interpretation of t is discussed in Section 128. We want a formula for the quantity t in the general case when we rank n objects, rather than 6. Since by definition, $t = S/(\max S)$, we have to find max S in terms of n. Now S takes on its maximum value when every pair contributes $+1$. There are $n(n - 1)/2$ pairs, so that

16.7
$$t = \frac{S}{\max S} = \frac{2S}{n(n-1)}.$$

EXAMPLE 16.8 In 1968 the members of the team that took first place in the American Baseball League had the following numbers of home runs (HR) and runs batted in (RBI):

Player	*A*	*B*	*C*	*D*	*E*	*F*	*G*	*H*
HR	15	16	1	12	21	11	36	25
RBI	63	56	12	37	90	60	85	84

What is the rank correlation coefficient between home runs and runs batted in?

We rearrange players according to number of home runs:

Player	*C*	*F*	*D*	*A*	*B*	*E*	*H*	*G*
HR	1	11	12	15	16	21	25	36
RBI	12	60	37	63	56	90	84	85

The quantity S is computed from the rearranged sequence of runs batted in,

$$S = (7 - 0) + (4 - 2) + (5 - 0) + (3 - 1) + (3 - 0) + (0 - 2) + (1 - 0) = 18.$$

It follows that

$$t = \frac{2 \times 18}{8 \times 7} = \frac{9}{14} = .64.$$

128* The Kendall rank correlation coefficient, unlike most other correlation coefficients, has a simple operational interpretation. To discover the meaning of t, we look at the computation of S in greater detail. We assume that neither x- nor y-observations contain ties to avoid unessential complications.

The first step in computing S consists in arranging x-observations from the smallest to the largest and then looking at pairs of y-observations. Let C equal the number of pairs with $y_i < y_j$ and let D equal the number of pairs with $y_i > y_j$, where always $i < j$. Then

$$S = C - D,$$

since a pair of the first type contributes $+1$ and a pair of the second type contributes -1 when computing S.

Let us now adopt the following terminology. Two pairs of observations (x_i, y_i) and (x_j, y_j) are said to form a *concordant* set if x- and y-observations change in the same direction. They are said to form a *discordant* set if x- and y-observations change in opposite directions. Thus a set is concordant if either

$$x_i < x_j \quad \text{and} \quad y_i < y_j \qquad \text{or} \qquad x_i > x_j \quad \text{and} \quad y_i > y_j.$$

A set is discordant if either

$$x_i < x_j \quad \text{and} \quad y_i > y_j \qquad \text{or} \qquad x_i > x_j \quad \text{and} \quad y_i < y_j.$$

(In Example 16.8, players B and F form a discordant set while players B and E form a concordant set.) With this new terminology, $C =$ #(concordant sets) and $D =$ #(discordant sets), so that

$$S = \#(\text{concordant sets}) - \#(\text{discordant sets}),$$

where #( ) stands for the number of items that have the property indicated within parentheses.

We now define two probabilities, p_c and p_d, that are called probabilities of concordance and discordance, respectively. These probabilities have the following interpretation. If we select at random two pairs of observations (x_1, y_1) and (x_2, y_2), the probability that the two pairs form a concordant set is p_c and the probability that they form a discordant set is p_d. Since our sample $(x_1, y_1), \ldots, (x_n, y_n)$ can be used to form $M = \frac{1}{2}n(n - 1)$ sets of which C are concordant and D are discordant, p_c is estimated as C/M while p_d is estimated as D/M.

It follows that the Kendall rank correlation coefficient

$$t = \frac{2S}{n(n-1)} = \frac{C}{M} - \frac{D}{M}$$

is an estimate of a parameter τ which is equal to the difference

$$\tau = p_c - p_d$$

of the probabilities of concordance and discordance.

A value of τ near 0 implies that x- and y-observations are just as likely to vary in the same direction as in opposite directions. A value of τ near $+1$ indicates that x- and y-observations are much more likely to vary in the same direction than in opposite directions; a value of τ near -1 implies just the opposite. The Kendall rank correlation coefficient compares the probabilities of concordance and discordance.

For the particular situation of two "judges" ranking n "contestants," τ, p_c, and p_d have still simpler interpretations. In such a case p_c is the probability that the two judges agree when ranking two randomly selected contestants; p_d is the probability that the two judges disagree; and $\tau = p_c - p_d$ measures the difference of the probabilities of agreement and disagreement.

129 *The Spearman Rank Correlation Coefficient*

The Kendall rank correlation coefficient is not the only way to measure strength of relationship between two rankings. Let the first ranking be $r_1, r_2, \ldots, r_n$ and the second ranking be $s_1, s_2, \ldots, s_n$. In Table 16.6, $r_1 = 4$, $r_2 = 6$, and so on, and $s_1 = 3$, $s_2 = 6$, and so on.

Two rankings are in complete agreement if $r_1 = s_1, r_2 = s_2, \ldots, r_n = s_n$. A natural way to measure deviation from complete agreement is to compute

$$D = (r_1 - s_1)^2 + (r_2 - s_2)^2 + \cdots + (r_n - s_n)^2.$$

If two rankings are in complete agreement, then $D = 0$. It can be shown that if one ranking is the reverse of the other, that is, if the two rankings are in complete disagreement, then $D = n(n^2 - 1)/3$. Thus D takes values between 0 and $n(n^2 - 1)/3$:

0 — D — $\frac{n(n^2-1)}{3}$

As a measure of strength of relationship of two rankings, D has a similar disadvantage as the quantity S: D has meaning only relative to the two extremes 0 and $n(n^2 - 1)/3$. We get around this inconvenience by considering instead the quantity

$$r_S = 1 - \frac{6D}{n(n^2-1)},$$

which is known as the Spearman rank correlation coefficient. If $D = 0$, implying identical rankings, then $r_S = +1$. If $D = n(n^2 - 1)/3$, implying reversed rankings, then $r_S = -1$. Like most correlation coefficients, r_S takes values between -1 and $+1$.

EXAMPLE 16.9 Find the Spearman rank correlation coefficient between the psychology and statistics grades in Table 16.4. We convert the psychology and statistics grades to ranks:

Psychology	9	1	7	5	8	6	2	4	10	3
Statistics	9	3	10	2	6	7	1	4	5	8

Then $D = (9 - 9)^2 + (1 - 3)^2 + \cdot \cdot \cdot + (3 - 8)^2 = 78$ and $r_S = 1 - (6 \times 78)/(10 \times 99) = .53$.

Operationally, the meaning of the Spearman rank correlation coefficient is considerably more complex than that of the Kendall rank correlation coefficient. Numerically, the Spearman coefficient is usually larger than the Kendall coefficient, but not invariably so.

Unless n is quite small, the hypothesis of independence of two variables x and y can be tested with the help of the statistic

$$z = \sqrt{n - 1}\, r_S.$$

For our example, $z = \sqrt{10 - 1}(.53) = 1.59$, whose two-sided descriptive level equals .11. This does not differ greatly from the result based on the Kendall rank correlation coefficient.

PROBLEMS

General instructions: Compute the Kendall rank correlation coefficient unless the Spearman coefficient is specifically called for.

1 For a sample of size 15, you have found that $S = -32$. Using a significance level of .05, test the hypothesis of randomness against the alternative of

a. a monotone trend;
b. a downward trend.

2 You have made eight successive determinations of the salinity of a body of water:

2.33 2.24 2.29 2.20 2.18 2.25 2.19 2.15

Would you say that the degree of salinity is decreasing over time?

3 Select ten, five-digit random numbers from Table J and test for randomness.

4 A local bookstore compared its best-seller list with the national best-seller list. Here are the results:

Position on National List	1	2	3	4	5	6	7
Position on Local List	1	3	2	6	4	7	5

Would you say that the local bookstore sales agree with national trends?

5 To study water pollution of a river, a scientist measured the concentration of a certain organic compound (y-values) and the amount of rainfall during previous weeks (x-values):

x	0.91	1.33	4.19	2.68	1.86	1.17
y	0.1	1.1	3.4	2.1	2.6	1.0

Is there a relationship between pollution level and amount of rainfall?

6 The following are the systolic and diastolic blood pressures (in millimeters) of ten persons:

Systolic	100	174	120	124	130	118	142	130	152	106
Diastolic	68	108	64	66	90	72	84	82	88	76

Can we conclude that systolic and diastolic blood pressure are related?

7 At a certain hospital the following number of babies were born between 1954 and 1973 (reading from left to right):

2406	2570	2549	2481	2419	2436	2453	2372	2213	2126
2143	1968	1960	1944	1884	1844	1806	1769	1726	1696

Test for randonmness. What is the descriptive level associated with the data?

8 Polls regularly estimate the national "approval rate" of the President's performance in office. The approval rate estimates for 1973 were as follows:

Date	*Percent*
January 15	51
February 1	68
February 19	65
April 2	59
April 9	54
April 30	49
June 11	44
August 6	31
August 18	38
September 24	32
October 8	30
October 22	29
November 5	27
December 3	31
December 10	29

Test for trend.

9 Find the rank correlation coefficient for the two rankings of critics 3 and 4 in Problem 6 of Chapter 15.

10 Two typists ranked five typewriters according to performance:

Typewriter	*A*	*B*	*C*	*D*	*E*
Typist 1	3	4	5	2	1
Typist 2	3	5	4	2	1

Compute the rank correlation coefficient for the two rankings.

11 Two judges at a figure skating contest assigned place numbers as follows to ten contestants:

Judge A	2	5	6	4	1	7	9	10	3	8
Judge B	1	4	5	6	2	7	10	8	3	9

a. Find the rank correlation coefficient.
b. Test for independence. What conclusion can we draw from the test result?

12 Find the rank correlation coefficient for the test scores for methods 1 and 2 in Problem 9 of Chapter 15. Do the same for methods 1 and 3. Would you say that the experimenter has been successful in arranging the 21 pupils in homogeneous groups?

13 Eight students applying for admission to a graduate school showed the following scores on the verbal and quantitative parts of the Graduate Record Examination (the first number in each pair represents the verbal score; the second, the quantitative score): (560, 620), (605, 490), (585, 552), (678, 632), (621, 512), (482, 780), (615, 665), (530, 545).

a. Find the rank correlation coefficient.
b. Test for independence.

14* Two contest judges ranked five contestants as follows:

Contestant	*A*	*B*	*C*	*D*	*E*
Judge 1	1	2.5	2.5	4	5
Judge 2	1	2.5	2.5	4	5

Show that $t = .9$. Note that even though the two judges used identical rankings, the value of t is less than $+1$. This is due to the fact that the two judges could not distinguish between contestants B and C.

15 Consider the data for Example 16.5.

a. If we leave out New York, are housing and transportation related?
b. How do you explain the difference between the present results and those arrived at in Example 16.5?

16 A testing organization determined the number of miles per gallon of gasoline for eight models of cars in city and highway driving. The results were as follows:

	Car Model							
	A	*B*	*C*	*D*	*E*	*F*	*G*	*H*
City Driving (m.p.g.)	26.2	24.8	23.6	22.0	20.1	19.6	18.4	16.7
Highway Driving (m.p.g.)	26.6	27.4	23.6	25.7	23.6	21.7	21.3	19.6

Find the rank correlation coefficient.

17 The sports editor of a student newspaper kept a record of the number of points scored and the number of fouls committed by each of the school's basketball players during a season. Here are the results:

Player	1	2	3	4	5	6	7	8	9	10	11	12
Points	1	0	3	2	20	10	55	49	60	57	45	72
Fouls	0	1	3	4	7	9	15	17	18	22	24	31

Find the rank correlation between number of fouls and points scored.

18 The following table gives the nicotine content (in milligrams) of thirty brands of cigarettes arranged according to increasing amounts of tar (in milligrams):

Tar	*Nicotine*	*Tar*	*Nicotine*
1	0.1	18	1.3
3	0.3	19	1.3
4	0.2	20	1.5
6	0.3	21	1.4
7	0.3	22	1.6
8	0.5	23	1.2
9	0.6	24	1.6
10	0.4	25	1.5
11	0.8	26	1.0
12	0.9	27	1.4
13	1.1	28	1.7
14	1.0	29	1.8
15	0.5	30	2.1
16	1.1	31	1.9
17	1.2	35	2.4

a. Make a scatter diagram for the data.
b. Find the rank correlation coefficient between amount of tar and nicotine content.
c. Test for independence.

19 Find the rank correlation coefficient between first digits and their frequency of appearance using the data in Problem 6 of Chapter 9.

20 Find the Spearman rank correlation coefficient using the data in

a. Problem 10;
b. Problem 11;
c. Problem 13;

d. Problem 16;
e. Problem 18.

21 In Table 16.2 arrange statistics grades according to size and compute S from the corresponding arrangement of psychology grades. How does the new S-value compare with the old S-value? Do you think this is a general result? Why?

22* According to Section 128, $S = C - D$, where $C = \#(\text{concordant sets})$ and $D = \#(\text{discordant sets})$ among the $\frac{1}{2}n(n - 1)$ sets of pairs (x_i, y_i) and (x_j, y_j). Find the values of C and D for Table 16.2 *without* first rearranging the data as in Table 16.4. This problem shows that we can compute S and t without first ordering x-observations according to size.

ONE-SAMPLE PROBLEMS FOR NORMAL POPULATIONS

130 In this chapter we start the study of *normal theory* methods. The name arises from the implicit assumption that sample observations follow a normal distribution. Sometimes the term *classical* methods is used, since these methods have been the primary working tools of statisticians for well over fifty years. The nonparametric methods of Chapters 12 to 16, with the exception of the sign test and Spearman rank correlation, are of more recent origin.

In Chapter 11 we stated our reasons for preferring the nonparametric approach to the classical approach. Recent statistical literature indicates that the nonparametric approach offers an important alternative to the classical approach. Nevertheless, in view of the widespread use of classical methods, it is appropriate to discuss some normal theory methods at this point. To keep the book within manageable bounds, the treatment will be relatively short and concise. The author hopes that the experience and insight that students have gained from studying nonparametric methods will help as they cope with the complexity of normal theory methods.

Point Estimates for the Parameters of a Normal Distribution†

131 A normal distribution is characterized by two parameters, the mean μ and the standard deviation σ. We start with point estimates of these two parameters.

†Students who are unfamiliar with summation notation should study Section 154 in the Appendix.

Let $x_1, \ldots, x_n$ represent a random sample from a normally distributed population. Statistical theory suggests the following point estimates of μ and σ:

17.1
$$\bar{x} = \frac{x_1 + \cdot \cdot \cdot + x_n}{n} = \frac{1}{n}\Sigma x$$

17.2 and
$$s = \sqrt{\frac{1}{n-1}\Sigma(x - \bar{x})^2}.$$

These estimates are called the *sample mean* and *sample standard deviation,* respectively.

The mean μ is the center of a normal distribution. The sample mean, $\bar{x}$, is the arithmetic mean of the sample observations and is used to estimate the population mean.

The standard deviation σ is a measure of variability. A large standard deviation indicates a population that is spread out. A small standard deviation indicates a population that is concentrated around the mean μ. There are many different ways to measure variability of a set of observations $x_1, \ldots, x_n$. In Problem 14 of Chapter 1 we suggested the *range,* the difference between the largest and the smallest of a set of numbers, as a possible measure. It is possible to estimate the population standard deviation σ of a normal distribution from the sample range. A better estimate, indeed the best possible estimate (see the discussion in Section 4 about comparing two estimates), is provided by the rather awkward looking sample standard deviation (17.2). The sample standard deviation measures variability by squaring the distances of each individual observation from the sample mean $\bar{x}$ and then compensating for squaring by taking the square root of an appropriate average.

To find s, we compute the sample variance

17.3
$$s^2 = \frac{1}{n-1}\Sigma(x - \bar{x})^2.$$

This is an estimate of the population variance σ^2. We will show that $\Sigma(x - \bar{x})^2$ can be computed as follows:

17.4
$$\Sigma(x - \bar{x})^2 = \Sigma x^2 - \frac{1}{n}(\Sigma x)^2.$$

EXAMPLE 17.5 Find the sample mean, variance, and standard deviation for the following five observations: 75, 82, 68, 87, 79.

We have

$$\Sigma x = 75 + 82 + 68 + 87 + 79 = 391$$

and
$$\Sigma x^2 = 75^2 + 82^2 + 68^2 + 87^2 + 79^2 = 30783$$

so that

$$\overline{x} = \frac{391}{5} = 78.2,$$

$$\Sigma(x - \overline{x})^2 = 30783 - \frac{391^2}{5} = 206.8,$$

$$s^2 = \frac{206.8}{4} = 51.7 \qquad \text{and} \qquad s = \sqrt{51.7} = 7.2.$$

The computing formula (17.4) is derived as follows:

$$\begin{aligned}\Sigma(x - \overline{x})^2 &= \Sigma(x^2 - 2\overline{x}x + \overline{x}^2) \\ &= \Sigma x^2 - 2\overline{x}\,\Sigma x + n\overline{x}^2 \\ &= \Sigma x^2 - \frac{2}{n}(\Sigma x)^2 + \frac{1}{n}(\Sigma x)^2 \\ &= \Sigma x^2 - \frac{1}{n}(\Sigma x)^2.\end{aligned}$$

132 *The Sampling Distribution of $\overline{x}$*

In Chapter 6, when we discussed estimation of the success probability p for binomial trials, we pointed out that point estimates by themselves are of little use without an indication of the expected sample fluctuation. For $\overline{x}$ such information is obtained from the following theorem (see Problem 20):

THEOREM 17.6 In random samples from normal populations with mean μ and standard deviation σ, the sample mean $\overline{x}$ is normally distributed with mean μ and standard deviation $\sigma/\sqrt{n}$.

Figure 5.29 shows that with probability γ a normally distributed variable deviates at most z_γ standard deviations from its mean μ, where z_γ-values are given in Table C. If we apply this result to the sample mean $\overline{x}$, we find that with probability γ the estimate $\overline{x}$ of μ deviates from μ by at most $z_\gamma\sigma/\sqrt{n}$. Thus the following inequality holds with probability γ:

17.7
$$\mu - z_\gamma \frac{\sigma}{\sqrt{n}} \leq \overline{x} \leq \mu + z_\gamma \frac{\sigma}{\sqrt{n}}.$$

$\overline{x}$

$\mu - z_\gamma\sigma/\sqrt{n}$ μ $\mu + z_\gamma\sigma/\sqrt{n}$

By taking a sufficiently large number of observations, we can make $\varepsilon = z_\gamma\sigma/\sqrt{n}$ as small as we please. As in the binomial case, a two-fold increase in accuracy (that is, a reduction of ε by a factor 2) is achieved by using four times as many observations, a ten-fold increase in accuracy, by using 100 times as many observations.

EXAMPLE 17.8 It is known that scores on an examination are normally distributed with standard deviation $\sigma = 10$. According to Table C, with probability $\gamma = .954$, the score of a single student may deviate from the population mean μ by as much as $z_\gamma \sigma = 2 \times 10 = 20$ points in either direction. If 25 students take the examination, with the same probability .954, their mean score $\bar{x}$ will not deviate from the population mean μ by more than $z_\gamma \sigma/\sqrt{n} = 2 \times 10/\sqrt{25} = 4$ points. Thus if we estimate the population mean μ on the basis of the scores of 25 randomly selected students, we can be fairly certain that the resulting estimate does not deviate from the true population mean by more than 4 points.

Confidence Intervals for the Mean of a Normal Distribution

133 The inequality (17.7) is easily converted into a confidence interval for the parameter μ. Whenever $\bar{x}$ does not deviate by more than $z_\gamma \sigma/\sqrt{n}$ from μ, μ does not deviate by more than $z_\gamma \sigma/\sqrt{n}$ from $\bar{x}$. Thus

17.9 $$\bar{x} - z_\gamma \sigma/\sqrt{n} \leq \mu \leq \bar{x} + z_\gamma \sigma/\sqrt{n}$$

is a confidence interval for μ which has confidence coefficient γ.

The confidence interval (17.9) requires knowledge of the true population standard deviation σ. In practice, the assumption that σ is known is rarely realistic. Thus we have the problem of what to do when σ is unknown or inaccurately known. For large samples the problem is solved by replacing the unknown σ in (17.9) by its sample estimate, s. Thus in large samples we use the following confidence interval,

17.10 $$\bar{x} - z_\gamma s/\sqrt{n} \leq \mu \leq \bar{x} + z_\gamma s/\sqrt{n}.$$

However, when n is smaller than 30 or 40, generally this solution is unsatisfactory, since the true confidence coefficient associated with the interval (17.10) is smaller than suggested by z_γ. For example, if $n = 6$ and we use $z_\gamma = 2$, corresponding to the value $\gamma = .954$ according to Table C, the true confidence coefficient associated with the interval (17.10) is .90.

We can adjust for this deficiency by replacing the quantity z_γ by a somewhat larger quantity t_γ, which is read from a table of the so-called t-distribution with $n - 1$ degrees of freedom. Appropriate t-values are found in Table H. Inspection of Table H shows that if n is as large as 40, there is very little difference between t_γ and z_γ (which is tabulated in Table H under ∞).

When sampling from a normal population with unknown standard deviation σ, we then use the following confidence interval for the population mean μ:

17.11 $$\bar{x} - t_\gamma s/\sqrt{n} \leq \mu \leq \bar{x} + t_\gamma s/\sqrt{n},$$

where t_γ has $n - 1$ degrees of freedom.

EXAMPLE 17.12 Assume that the five observations in Example 17.5 have come from a normal population with unknown mean μ. We want to find a confidence interval with confidence coefficient .95 for μ. Using the results of Example 17.5, we find the interval:

$$78.2 - 2.776 \times 7.2/\sqrt{5} \leq \mu \leq 78.2 + 2.776 \times 7.2/\sqrt{5}$$

or $$69.3 \leq \mu \leq 87.1.$$

We have remarked repeatedly that the length of a confidence interval is a good indicator of its usefulness. The confidence interval (17.11) for the mean of a normal distribution whose standard deviation is unknown has length $2t_\gamma s/\sqrt{n}$. The length of the interval depends on the sample standard deviation s and therefore varies from sample to sample. Since the sample standard deviation s estimates the population standard deviation σ, we can estimate the length of the confidence interval only if we have some idea of the magnitude of σ. Experimenters often have some rough knowledge. Aside from this indeterminacy, the length of the interval depends on the confidence coefficient γ and the sample size n. As on earlier occasions, the length of the interval increases together with γ and decreases inversely as the square root of n. We require a four-fold increase in the number of observations to achieve a reduction in length by a factor 2.

Tests of Hypotheses about the Mean of a Normal Distribution

134 The hypothesis $\mu = \mu_0$ can be tested against the two-sided alternative $\mu \neq \mu_0$ at significance level $\alpha'' = 1 - \gamma$ by determining if the hypothetical value μ_0 is in the confidence interval (17.11).

Alternatively, according to Theorem 17.6, the quantity

17.13 $$z = \frac{\bar{x} - \mu}{\sigma/\sqrt{n}}$$

has a standard normal distribution. If we replace the population standard deviation σ by its sample estimate s in (17.13), the resulting quantity

17.14 $$t = \frac{\bar{x} - \mu}{s/\sqrt{n}}$$

has the t-distribution with $n - 1$ degrees of freedom. A test of the hypothesis $\mu = \mu_0$ against both one- and two-sided alternatives proceeds as follows. We replace the parameter μ in (17.14) by the hypothetical value μ_0 and reject the hypothesis under consideration when the resulting t-value falls in the critical region according to Table H for $n - 1$ degrees of freedom.

EXAMPLE 17.15 We want to test the hypothesis $\mu = 70$ at significance level .05 using the assumptions of Example 17.12. We find that

$$t = \frac{78.2 - 70}{7.2/\sqrt{5}} = 2.55.$$

The two-sided critical region consists of t-values less than -2.776 or greater than 2.776. The computed value $t = 2.55$ is not in the critical

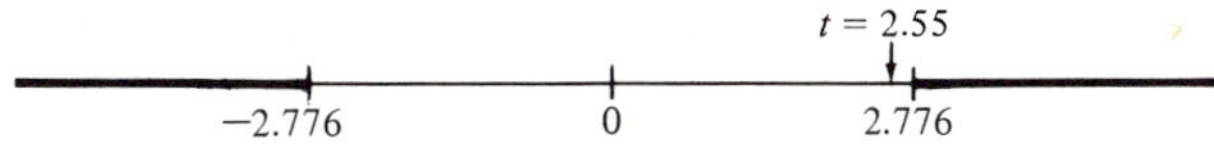

region. So we do not reject the hypothesis $\mu = 70$ against the two-sided alternative $\mu \neq 70$. Against the one-sided alternative $\mu > 70$, the critical region consists of sufficiently large values of t, namely $t > 2.132$. The computed t-value indicates rejection:

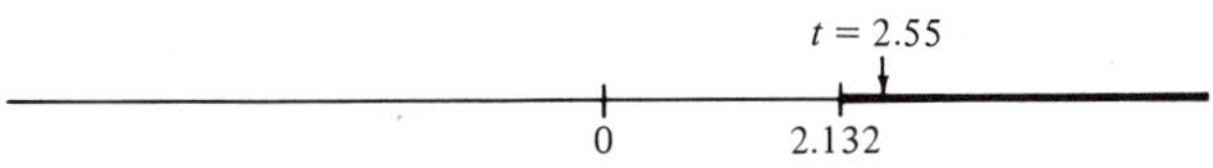

Inspection of Table H for four degrees of freedom shows that the one-sided descriptive level associated with the observed value $t = 2.55$ is between .025 and .05.

The Central Limit Theorem

135 The procedures discussed so far in this chapter are based on the theorem that in random samples from a normal population with mean μ and standard deviation σ, the sample mean $\bar{x}$ is normally distributed with mean μ and standard deviation $\sigma/\sqrt{n}$. One of the most remarkable and useful theorems of the theory of probability, the Central Limit Theorem, states that this result remains approximately true for random samples from (almost) any kind of population, provided that the number of observations n in the sample is sufficiently large. In large samples, we do not even have to assume that the population standard deviation is known. Using the sample standard deviation s in place of the population standard deviation σ usually provides satisfactory results. It then follows that (17.10) provides a confidence interval for the mean μ of an arbitrary population, provided the number of observations is sufficiently large. The hypothesis $\mu = \mu_0$ is tested by computing the statistic

17.16 $$z = \frac{\bar{x} - \mu_0}{s/\sqrt{n}} = \frac{\sqrt{n}(\bar{x} - \mu_0)}{s}.$$

We accept or reject the hypothesis according to the appropriate critical values in Table C. Alternatively, (17.16) can serve to compute the descriptive level associated with a given experimental result. The great popularity and usefulness of normal theory methods are due largely to the wide applicability of the Central Limit Theorem.

There is no simple answer to the question how large is sufficiently large to justify recourse to the Central Limit Theorem, but few statisticians will hesitate to use (17.16) if n is at least 40 or 50.

EXAMPLE 17.17 A traveling salesperson keeps an expense record by rounding off each entry to the nearest dollar. Thus if a meal costs \$4.65, it is recorded as \$5 for a rounding error of $5 - 4.65 = .35$. When a gasoline charge is \$6.30, it is recorded as \$6 for a rounding error of $-.30$. The salesperson is curious about the total rounding error for a week's record.

The salesperson keeps an exact record and finds that in a week in which there are 36 entries, there is a total error of \$1.48. The standard deviation of the 36 entries is .28. Using (17.10), we can find the following confidence interval ($\gamma = .95$) for the mean error:

$$\frac{1.48}{36} - \frac{1.96 \times .28}{\sqrt{36}} \leq \mu \leq \frac{1.48}{36} + \frac{1.96 \times .28}{\sqrt{36}}$$

or

$$-.05 \leq \mu \leq .13.$$

We get a confidence interval for the mean weekly total error (based on 36 entries) by multiplying by 36. Thus

$$-\$1.80 \leq \text{mean weekly error} \leq \$4.68.$$

Inference Procedures for the Standard Deviation

136* *A Confidence Interval for σ*

It can be shown that, in random samples of n observations from a normal distribution with standard deviation σ, the quantity

$$\frac{(n-1)s^2}{\sigma^2} = \frac{1}{\sigma^2}\Sigma(x-\bar{x})^2$$

has a chi-square distribution with $n - 1$ degrees of freedom. This result can be used to find a confidence interval for σ having confidence coefficient γ as follows. Let $\alpha = 1 - \gamma$. From a table of the chi-square distribution, we find two values V_1 and V_2 such that the area to the left

FIGURE 17.18

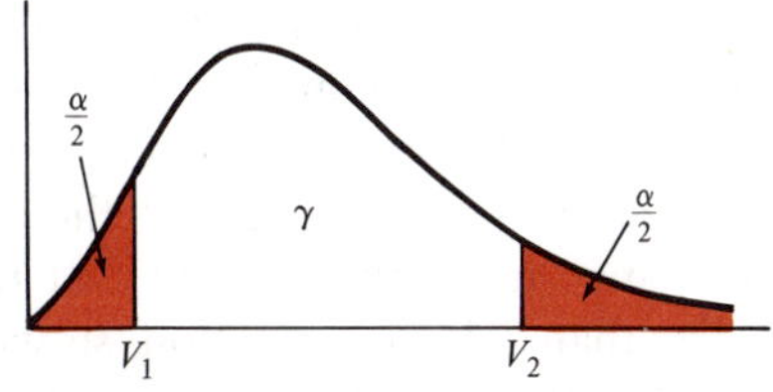

of V_1 is $\alpha/2$ and the area to the right of V_2 is also $\alpha/2$. Then with probability γ,

$$V_1 \leq \frac{\Sigma(x-\bar{x})^2}{\sigma^2} \leq V_2,$$

which can also be written as

$$\frac{1}{V_2} \leq \frac{\sigma^2}{\Sigma(x-\bar{x})^2} \leq \frac{1}{V_1}.$$

Multiplying all three parts by $\Sigma(x - \bar{x})^2$, we find the confidence interval for σ^2,

$$\frac{\Sigma(x-\bar{x})^2}{V_2} \leq \sigma^2 \leq \frac{\Sigma(x-\bar{x})^2}{V_1}.$$

But then

17.19 $$\sqrt{\frac{\Sigma(x-\bar{x})^2}{V_2}} \leq \sigma \leq \sqrt{\frac{\Sigma(x-\bar{x})^2}{V_1}}$$

is a confidence interval for σ.

EXAMPLE 17.20 We want to find a confidence interval for σ using the data of Example 17.5. We find $V_1 = .484$ and $V_2 = 11.1$ corresponding to the confidence coefficient .95. Thus the lower confidence bound for σ^2 is $206.8/11.1 = 18.6$; the upper bound is $206.8/.484 = 427.3$. The corresponding bounds for σ are $\sqrt{18.6} = 4.3$ and $\sqrt{427.3} = 20.7$.

137* *Testing a Hypothesis about σ*

The confidence interval in (17.19) provides a two-sided test of the hypothesis $\sigma = \sigma_0$. Alternatively, this hypothesis can be tested by referring the statistic $\Sigma(x - \bar{x})^2/\sigma^2_0$ directly to a table of the chi-square distribution with $n - 1$ degrees of freedom.

EXAMPLE 17.21 We want to test the hypothesis $\sigma = 10$ using the data of Example 17.5. The appropriate test statistic is $\Sigma(x - \bar{x})^2/\sigma^2_0 = 206.8/10^2 = 2.068$. Suppose that we have decided on a significance level .05. When testing against the two-sided alternative $\sigma \neq 10$, we accept the null hypothesis, since the test statistic has a value between .484 and 11.1. This decision agrees of course with the confidence interval of Example 17.20. As a second illustration, consider the case where we should like to find out whether the true standard deviation is less than 10. We would then test the hypothesis $\sigma \geq 10$ against the alternative $\sigma < 10$. Using again a significance level .05, we find that the test statistic would have to be smaller than .711 for rejection, so we accept the hypothesis. It would seem that the true standard deviation is at least 10.

In solving Examples 17.20 and 17.21, we have assumed that the five observations represent a random sample from some normal population.

PROBLEMS

All problems should be solved using the methods of this chapter. The student should make any appropriate assumptions.

1 Find the mean and standard deviation for the following six observations:

3 1 9 4 8 5

2 For ten observations, you have computed $\Sigma x = 495$, $\Sigma x^2 = 27611$.

a. Find $\bar{x}$.
b. Find s.

3 Find the sample mean and standard deviation for the differences in (13.1).

4 Find the sample mean and standard deviation for the turn-around times in Problem 5 of Chapter 12. (*Note:* The mean and standard deviation are important sample characteristics for observations that are not necessarily normally distributed. There are strong indications that the present sample does not come from a normally distributed population. Explain.)

5 Find the sample mean for the data of

a. Example 12.4;
b. Table 12.11;
c. Problem 9 of Chapter 12.

In each case compare the mean with the median for the same set of data. How do you explain that the mean and median are very similar in some cases, while, in others, they are not?

6 Find the sample standard deviation for the data of

a. Example 12.4;
b. Table 12.11;
c. Problem 9 of Chapter 12.

7 To what extent does the test result in Example 17.15 agree with the confidence interval in Example 17.12?

8 One hundred newly minted pennies were weighed to the nearest tenth of a milligram. The sum of the measurements equaled 3110; the sum of the squares of the measurements equaled 96740. Let μ be the mean weight of newly minted pennies.

a. Find a point estimate for μ.
b. Find a confidence interval for μ.

9 The first ten pennies in Problem 8 weighed

31.0 31.4 30.4 30.1 30.6 31.1 31.2 30.9 30.3 30.8.

Using only this sample information,
a. find a confidence interval for the mean weight of newly minted pennies;
b. test the hypothesis $\mu = 30$ against the alternative $\mu \neq 30$;
*c. test the hypothesis $\sigma = .4$.

10 Chapter 12 is concerned with the solution of one-sample problems. Find three problems in the problem set for Chapter 12 that should not be solved using the methods of Chapter 17.

11 In Problem 2 of Chapter 12, test the hypothesis that the mean score for the test is 75.

12 In Problem 4 of Chapter 12, find

a. a point estimate for the mean fat content;
b. a confidence interval for the mean fat content.

13 Solve Problem 9 of Chapter 12.

14 In Problem 15 of Chapter 12,

a. test the hypothesis that the mean diastolic blood pressure equals 80;
b. find a confidence interval ($\gamma = .95$) for the mean diastolic blood pressure.

15 In Problem 16 of Chapter 12, test the hypothesis that the mean score for students from College C equals 110.

16 In Problem 17 of Chapter 12,

a. find a point estimate for the mean weight gain;
b. find a confidence interval for the mean weight gain;
c. test the hypothesis that the mean weight gain is 70.

17* For the tire data in Problem 9 of Chapter 12,

a. test the hypothesis that $\sigma = 2500$;
b. find a confidence interval for σ.

18* Let $x_1, \ldots, x_n$ be a sample of x-values and let c be an arbitrary constant. Define a new sample of y-values by subtracting c from each x-value:

$$y_1 = x_1 - c, \ldots, y_n = x_n - c.$$

Show that $\bar{y} = \bar{x} - c$ and $s_y^2 = s_x^2$, where s_x^2 and s_y^2 are the sample variances of x- and y-values, respectively.

19* The computational work for computing the sample variance can often be reduced with the help of Problem 18. We make a rough guess of the mean of the sample observations and use this rough guess as the constant c in Problem 18. The sample variance of the original x-values equals the sample variance of the new y-values. The new y-values are often much more manageable than the original x-values. In Example 17.5 choose $c = 80$ and recompute the sample variance.

20* Let $X_1, \ldots, X_n$ be n independent random variables, each having mean μ and variance σ^2. Let $\bar{X} = (X_1 + \cdots + X_n)/n$.

a. Show that the mean of $\bar{X}$ is μ.
b. Show that the variance of $\bar{X}$ is σ^2/n.

(*Hint:* Study Example A.13. Use Theorem A.7 and Problem 10 of the Appendix.)

21 Problem 5 of Chapter 1 provides information on the number of children of 100 families in a certain community. In Chapter 1 you were asked to find the median number of children per family. Can you find the mean number of children?

18

TWO-SAMPLE PROBLEMS FOR NORMAL POPULATIONS

138 In this chapter we develop normal theory solutions for the type of problems discussed in Chapters 13 and 14. When samples from two normal populations are available, the investigator is most often interested in the difference of means $\Delta = \mu_x - \mu_y$.

The parameter Δ (see Figure 18.1) corresponds to the shift parameter Δ of Section 110 in Chapter 14. We will discuss point and interval estimates of Δ for normally distributed populations.

FIGURE 18.1

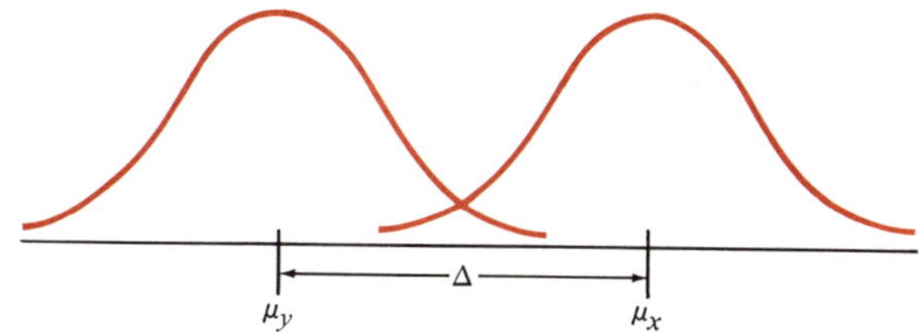

In Chapters 13 and 14 we discussed tests of the hypothesis that two samples—an x- and a y-sample—have come from one and the same population. In terms of the shift parameter Δ, this hypothesis states that $\Delta = 0$. Correspondingly, in the normal theory case we are frequently interested in the hypothesis

$$H: \quad \Delta = \mu_x - \mu_y = 0 \quad \text{or} \quad \mu_x = \mu_y.$$

The hypothesis H states that two normal populations have the same mean.

The following theorem is needed for our discussion:

THEOREM 18.2 If x and y are two normally distributed variables with means μ_x and μ_y and standard deviations σ_x and σ_y, then the difference $x - y$ is again normally distributed with mean $\mu_x - \mu_y$. If, in addition, x and y are independent, then the standard deviation of $x - y$ equals $\sqrt{\sigma_x^2 + \sigma_y^2}$.

Paired Observations

139 We first take up the case discussed in Chapter 13, when x- and y- observations occur in natural pairs: $x_1, y_1; x_2, y_2; \ldots; x_n, y_n$. As before, we consider the differences $d_1 = x_1 - y_1$, $d_2 = x_2 - y_2, \ldots,$ $d_n = x_n - y_n$. Theorem 18.2 implies that the set of differences $d_1, d_2,$ $\ldots, d_n$ represents a random sample from a normal population with mean $\Delta = \mu_x - \mu_y$. (It is usually appropriate to consider the standard deviation σ_d unknown.) We have reduced the two-sample problem to a one-sample problem, which can be solved by the methods of Chapter 17.

By (17.1) a point estimate of Δ is

$$\bar{d} = \frac{1}{n}\Sigma d.$$

By (17.11) a confidence interval for Δ is

$$\bar{d} - t_\gamma s_d/\sqrt{n} \leq \Delta \leq \bar{d} + t_\gamma s_d/\sqrt{n},$$

where by (17.2)

$$s_d = \sqrt{\frac{1}{n-1}\Sigma(d - \bar{d})^2}.$$

Finally, by (17.14) the hypothesis $\Delta = \Delta_0$ is tested by means of the t-statistic with $n - 1$ degrees of freedom,

$$t = \frac{(\bar{d} - \Delta_0)}{s_d/\sqrt{n}}.$$

In particular, the test of the hypothesis $\mu_x = \mu_y$ uses the statistic

$$t = \frac{\sqrt{n}\,\bar{d}}{s_d}.$$

EXAMPLE 18.3 In the suntan lotion experiment of Chapter 13, we observed the following differences: 5, 3, −1, 14, 7, −4, 11, 10. We will analyze the given data again using the normal theory approach we have just developed.

We find that $\bar{d} = 45/8 = 5.625$. Since the accuracy of the original observations is to the nearest integer, it is inappropriate to keep more than one decimal place when estimating Δ on the basis of only eight observations. Thus our estimate is 5.6.

For a 95 percent confidence interval, we have $t_\gamma s_d/\sqrt{n} = 2.365 \times 6.140/\sqrt{8} = 5.1$ so that

$$5.6 - 5.1 \leq \Delta \leq 5.6 + 5.1$$

or

$$0.5 \leq \Delta \leq 10.7.$$

Finally, for the test of the hypothesis $\Delta = \mu_x - \mu_y = 0$, we find that $t = \sqrt{8} \times 5.625/6.140 = 2.59$, indicating rejection of the hypothesis at significance level .05. The same result is implied by the confidence interval.

A Generalization

140 In Section 139 we were able to apply the methods of Chapter 17 simply by replacing the sample observations $x_1, \ldots, x_n$ of Chapter 17 by the sample differences $d_1, \ldots, d_n$ of Section 139. The problems that we want to discuss in the remainder of this chapter and in Chapter 20 are more complex. Nevertheless, it turns out that much of the work of Chapter 17 can serve as a model by the simple expedient of phrasing the basic problem of Chapter 17 in more general terms.

Let us review the pertinent facts of Chapter 17. We are interested in the parameter μ. Using sample observations, we can compute the point estimate $\bar{x}$ of μ. From general theory it is known that the point estimate $\bar{x}$ is normally distributed with mean μ and standard deviation $\sigma/\sqrt{n}$, where the population standard deviation σ can be estimated from sample observations by means of a quantity s. It then follows that the statistic

$$t = \frac{\bar{x} - \mu}{s/\sqrt{n}}$$

has a t-distribution. The appropriate number of degrees of freedom depends on the estimate s of σ. Hypotheses about μ are tested using the above statistic t, and a confidence interval for μ takes the form

$$\bar{x} - t_\gamma s/\sqrt{n} \leq \mu \leq \bar{x} + t_\gamma s/\sqrt{n}.$$

Our generalization consists in the following. We are interested in a parameter θ. Using sample observations, we can compute a point estimate e of θ. From general theory it is known that e is normally distributed with mean θ and standard deviation $c\sigma$, where c is a known constant and where the population standard deviation σ can be estimated from sample observations by means of a quantity s. (We are using the same symbol s to denote the estimate of σ, even though in general s will differ from one problem to another.) It follows that the statistic

18.4 $$t = \frac{e - \theta}{cs}$$

has a t-distribution with degrees of freedom equal to the number of degrees of freedom associated with the appropriate estimate s of σ. Hypotheses about θ are tested with the help of the statistic (18.4) and a confidence interval for θ has the form

18.5 $$e - t_\gamma cs \leq \theta \leq e + t_\gamma cs.$$

Independent Samples

141 When developing normal alternatives for the methods of Chapter 14, we assume that the observations $x_1, \ldots, x_m$ represent a random sample from a normal population with mean μ_x and standard deviation σ_x and that the observations $y_1, \ldots, y_n$, represent an independent random sample from a normal population with mean μ_y and standard deviation σ_y.

Experience shows that even though two treatments may differ in their effectiveness, that is, may have different means μ_x and μ_y, they often show approximately the same degree of variation from one observation to another, that is, have $\sigma_x = \sigma_y$. The classical two-sample problem assumes that we have independent samples from two normal populations with equal, but unknown, standard deviation $\sigma_x = \sigma_y = \sigma$ and are interested in the difference of means $\Delta = \mu_x - \mu_y$. For normal populations, this model coincides exactly with the shift model of Section 110 in Chapter 14.

A point estimate of Δ is the difference of sample means $\bar{x} - \bar{y}$. To find a confidence interval for Δ and derive a test statistic for tests involving Δ, we make use of the approach suggested in Section 140. We have $\theta = \Delta = \mu_x - \mu_y$ and $e = \bar{x} - \bar{y}$. Theorem 18.1 gives the remaining information. Using $\bar{x}$ in place of x and $\bar{y}$ in place of y, we find that $\bar{x} - \bar{y}$ is normally distributed with mean $\mu_x - \mu_y = \Delta$ and, in view of Theorem 17.6, with standard deviation

$$\sqrt{\frac{\sigma_x^{\,2}}{m} + \frac{\sigma_y^{\,2}}{n}} = \sqrt{\frac{\sigma^2}{m} + \frac{\sigma^2}{n}} = \sigma\sqrt{\frac{1}{m} + \frac{1}{n}},$$

so that $c = \sqrt{1/m + 1/n}$. We generalize 17.2, and use the following estimate of the common population standard deviation,

18.6 $$s = \sqrt{\frac{\Sigma(x - \bar{x})^2 + \Sigma(y - \bar{y})^2}{m + n - 2}},$$

which has $(m - 1) + (n - 1) = m + n - 2$ degrees of freedom. The test statistic (18.4) then becomes

18.7 $$t = \frac{(\bar{x} - \bar{y}) - \Delta_0}{s\sqrt{1/m + 1/n}},$$

which has a t-distribution with $m + n - 2$ degrees of freedom. The confidence interval (18.5) for Δ becomes

18.8 $$(\bar{x} - \bar{y}) - t_\gamma s\sqrt{1/m + 1/n} \le \Delta \le (\bar{x} - \bar{y}) + t_\gamma s\sqrt{1/m + 1/n}.$$

EXAMPLE 18.9 Let us consider again the two sets of examination scores in Table 14.1:

x-sample:	72	79	93	91	70	95	82	80	74	86	
y-sample:	78	66	65	84	69	73	71	75	68	90	76

Past experience may have told us that it is appropriate to consider test scores to be normally distributed with the same standard deviation in both cases. We can then use statistic (18.7) to test the hypothesis $\Delta = 0$. We compute $\Sigma(x - \bar{x})^2 = 707.6$ and $\Sigma(y - \bar{y})^2 = 592.9$, so that $s = \sqrt{(707.6 + 592.9)/19} = 8.27$. Then

$$t = \frac{82.2 - 74.1}{8.27\sqrt{1/10 + 1/11}} = 2.24.$$

With 19 degrees of freedom, the result is significant at the .05 level but not at the .01 level. For a 99 percent confidence interval, we find that $t_\gamma s\sqrt{1/m + 1/n} = 2.86 \times 8.27\sqrt{1/10 + 1/11} = 10.3$, giving the confidence interval

$$8.1 - 10.3 \le \Delta \le 8.1 + 10.3$$

or

$$-2.2 \le \Delta \le 18.4.$$

142* In Section 141 we have made the assumption that the two population standard deviations σ_x and σ_y are equal (or at least nearly equal). Violation of this assumption poses both practical and theoretical problems. From the practical viewpoint, if there are large differences in variation for the two populations, the investigator may want to reconsider the decision to look solely at the parameter Δ. Usually it will be much more appropriate to consider the consequences of any difference in population means in conjunction with the consequences of a difference in population variation.

Even if we should decide that $\Delta = \mu_x - \mu_y$ is of primary interest, substantial differences between σ_x and σ_y introduce theoretical problems. As long as the two sample sizes m and n are nearly equal, the methods of Section 141 provide rather reliable guidance. However, when m and n differ widely, the methods are no longer dependable. Many procedures to cover this case have been suggested in the statistical literature, but none is completely satisfactory. We mention briefly one possible procedure.

Whether $\sigma_x = \sigma_y$ or not, $\bar{x} - \bar{y}$ is always an appropriate point estimate of Δ. From Theorem 18.2 we find that $\bar{x} - \bar{y}$ has a normal distribution with mean $\mu_x - \mu_y = \Delta$ and standard deviation $\sqrt{\sigma_x^2/m + \sigma_y^2/n}$. Our customary approach of estimating an unknown population variance suggests consideration of the statistic

$$t^* = \frac{\bar{x} - \bar{y} - \Delta}{\sqrt{s_x^2/m + s_y^2/n}},$$

where

$$s_x^2 = \frac{1}{m-1}\Sigma(x - \bar{x})^2 \qquad \text{and} \qquad s_y^2 = \frac{1}{n-1}\Sigma(y - \bar{y})^2.$$

Theoretical problems arise from the fact that the exact distribution of t^* (and similar statistics) depends on the unknown ratio of population standard deviations σ_x/σ_y. If both m and n are large, this presents no problem, since t^* is approximately normally distributed and may simply be referred to Table C. When both m and n are small, one simple solution is to treat t^* as a variable having a t-distribution with $k - 1$ degrees of freedom, where k equals the smaller of m and n.

PROBLEMS

All problems should be solved using the methods of this chapter. For each problem the assumptions made by the student should be stated.

1 Show that $\bar{d} = \bar{x} - \bar{y}$.

2 Solve Problem 2 of Chapter 13.

3 Solve Problem 4 of Chapter 13.

4 In Problem 6 of Chapter 13, find a confidence interval for the mean increase in pulse rate due to exercise.

5 In Problem 7 of Chapter 13,

a. determine whether there is any difference in the mean number of hours of additional sleep for the two drugs;
b. find a point estimate for the mean difference;
c. find a confidence interval for the mean difference.

6 In Problem 8 of Chapter 13,

a. find a point estimate for the mean difference in gasoline mileage for the two types of tires;
b. find a confidence interval for the mean difference.

7 For the data of Problem 16 of Chapter 14, compute the value of the appropriate t-statistic. How many degrees of freedom does t have? Answer the question of the original problem.

8 In a sample of size 16 from a normal population, you have found $\Sigma x = 1205$ and $\Sigma x^2 = 92411$. In an independent sample of size 22 from another normal population, you have found $\Sigma y = 1767$ and $\Sigma y^2 = 143921$.

a. Test the hypothesis $\mu_x = \mu_y$.
b. Find a confidence interval for $\Delta = \mu_x - \mu_y$ having confidence coefficient .98.

9 The GRE scores of 10 students in Department X are: 646, 655, 582, 504, 407, 779, 696, 830, 772, 745. The GRE scores of 12 students in Department Y are: 649, 837, 690, 669, 634, 698, 548, 615, 685, 659, 657, 805. Would you say that the two departments differ in their admission policy? Support your statements by appropriate statistical arguments.

10 Using the data in Problem 21 of Chapter 12, test the hypothesis that male and female drivers obtain identical mileage from a set of tires.

11 In Problem 6 of Chapter 14, find a confidence interval for the difference of means of gasoline mileage for manual and automatic shift cars.

12 Do Problem 7 of Chapter 14.

13 Do Problem 8 of Chapter 14.

14 Do Problem 9 of Chapter 14.

19

k-SAMPLE PROBLEMS FOR NORMAL POPULATIONS

143 The problem to be considered in this chapter can be stated as follows. We have k independent samples, one each from k normal populations with means $\mu_1, \ldots, \mu_k$, and we want to test the hypothesis

$$H: \quad \mu_1 = \cdot \cdot \cdot = \mu_k.$$

Throughout this chapter we assume that the k normal populations have the same (unknown) variance σ^2. Violation of this assumption may seriously invalidate the results.

We find a test statistic for H by generalizing the test statistic (18.7) for the corresponding two-sample problem. In (18.7) the difference $\bar{x} - \bar{y}$ is a measure of how far apart the population means μ_x and μ_y are. Since $\bar{x}$ and $\bar{y}$ are subject to sampling fluctuations, we need a standard with which to compare the difference $\bar{x} - \bar{y}$. The denominator in (18.7) serves as such a standard. Similarly, in the present case we find a quantity (to be called MS_A) that is a measure of how far apart the population means $\mu_1, \ldots, \mu_k$ are. Again, because of sampling fluctuations, this measure has to be compared with an appropriate standard (to be called MS_W). We first indicate how the two quantities MS_A and MS_W are computed and how they are used to test the hypothesis H. An explanation of the various steps is postponed until Section 144.

We use the subscript j to enumerate the k samples. Thus j takes the possible values $1, 2, \ldots, k$. Let n_j be the number of observations in the jth sample and T_j, the sum of the n_j observations. By N we denote the total number of observations in all k samples,

$$N = n_1 + n_2 + \cdot \cdot \cdot + n_k;$$

by T, the sum of all N observations,

$$T = T_1 + T_2 + \cdot \cdot \cdot + T_k.$$

To find MS_A and MS_W, we compute three quantities:

$$C = \frac{T^2}{N},$$

$D =$ the sum of the squares of the N individual observations,

$$E = \frac{T_1^{\,2}}{n_1} + \cdot \cdot \cdot + \frac{T_k^{\,2}}{n_k}.$$

Then

$$MS_A = \frac{E - C}{k - 1}$$

and

$$MS_W = \frac{D - E}{N - k}.$$

We reject the hypothesis H provided that

$$F = \frac{MS_A}{MS_W}$$

is sufficiently large, that is, surpasses the appropriate critical value listed in Table I. To use Table I, we choose one of the following significance levels, $\alpha = .10$, .05, or .01. On the appropriate page, the critical value is listed at the intersection of the row labeled $N - k$ and the column labeled $k - 1$.

EXAMPLE 19.1 As an illustration of the required computations, we look once more at the data in Table 15.3 representing stopping distances for five brands of automobile tires. For the sake of convenience, the data are repeated in Table 19.2. The present analysis assumes that the data represent samples from normal populations having identical variance. We want to test the hypothesis

TABLE 19.2

Brand of Tire				
A	*B*	*C*	*D*	*E*
151	157	135	147	146
143	158	146	174	171
159	150	142	179	167
152	142	129	163	145
156	140	139	148	147
			165	166

$$\mu_A = \mu_B = \mu_C = \mu_D = \mu_E.$$

Computations proceed as follows:

	A	*B*	*C*	*D*	*E*	
T_j	761	747	691	976	942	$T = 4117$
n_j	5	5	5	6	6	$N = 27$

$$C = \frac{4117^2}{27} = 627766.3,$$

$$D = 151^2 + \cdot \cdot \cdot + 166^2 = 631775,$$

$$E = \frac{761^2}{5} + \cdot \cdot \cdot + \frac{942^2}{6} = 629578.9,$$

$$MS_A = \frac{629578.9 - 627766.3}{4} = 453.2,$$

$$MS_W = \frac{631775 - 629578.9}{22} = 99.8,$$

and
$$F = \frac{453.2}{99.8} = 4.54.$$

For $\alpha = .01$, we find the critical value 4.31 at the intersection of the row labeled 22 and the column labeled 4. Since $F = 4.54 > 4.31$, the hypothesis H is rejected at significance level .01.

144* *Analysis of Variance*

A precise and somewhat elaborate notation is helpful if we want to gain insight into the manipulations of Section 143. For the ith observation in the jth sample, we write

$$x_{ij}, \qquad i = 1, \ldots, n_j; \qquad j = 1, \ldots, k.$$

We have: the sum of the observations in the jth sample,

$$T_j = \Sigma_i x_{ij};$$

the mean of the observations in the jth sample,

$$\bar{x}_j = \frac{T_j}{n_j};$$

the sum of all N observations,

$$T = \Sigma_j T = \Sigma_i \Sigma_j x_{ij};$$

and the overall sample mean,

$$\bar{x} = \frac{T}{N}.$$

The analysis of variance receives its name from the fact that the so-called total sum of squares

$$SS_T = \Sigma_j \Sigma_i (x_{ij} - \bar{x})^2,$$

which characterizes the spread of all N observations about the overall mean $\bar{x}$, is written as the sum of two sums of squares representing different sources of variation. We write

$$x_{ij} - \bar{x} = (x_{ij} - \bar{x}_i) + (\bar{x}_i - \bar{x});$$

that is, we express the deviation of x_{ij} from the overall mean $\bar{x}$ as the sum of two deviations: (i) the deviation of x_{ij} from the mean $\bar{x}_j$ of all observations in the jth sample, and (ii) the deviation of the jth sample mean from the overall mean $\bar{x}$. Substitution into SS_T and some lengthy algebra give

$$SS_T = SS_W + SS_A,$$

where $$SS_W = \Sigma_j \Sigma_i (x_{ij} - \bar{x}_j)^2 = D - E$$

and $$SS_A = \Sigma_j n_j (\bar{x}_j - \bar{x})^2 = E - C$$

Since for each j, $\Sigma_i (x_{ij} - \bar{x}_j)^2$ measures variability within the jth sample, SS_W is called the within sum of squares, while SS_A is usually called the among or treatment sum of squares.

Associated with each sum of squares is an appropriate number of degrees of freedom: $N - 1$ for SS_T, $\Sigma_j (n_j - 1) = N - k$ for SS_W, $(N - 1) - (N - k) = (k - 1)$ for SS_A. By dividing SS_W and SS_A by the respective number of degrees of freedom, we obtain the *mean squares*

$$MS_W = \frac{SS_W}{(N - k)} \quad \text{and} \quad MS_A = \frac{SS_A}{(k - 1)}.$$

These mean squares have the following significance for our problem. MS_W is an estimate of the unknown variance σ^2 common to all k normal populations, whether H is true or not. On the other hand, the treatment mean square MS_A is an estimate of

$$\sigma_A{}^2 = \sigma^2 + \frac{1}{k - 1} \Sigma_j n_j (\mu_j - \mu)^2,$$

where $\mu = \Sigma_j n_j \mu_j / N$. Thus MS_W measures fluctuation caused by sampling, while MS_A measures also how far apart $\mu_1, \ldots, \mu_k$ are.

When the hypothesis H is true, that is, when $\mu_1 = \cdots = \mu_k = \mu$, then $\sigma_A{}^2 = \sigma^2$. When the hypothesis H is false, we have $\sigma_A{}^2 > \sigma^2$. When we compute the test statistic $F = MS_A / MS_W$, we are trying to decide whether $\sigma_A{}^2 = \sigma^2$, indicating that H is true, or whether $\sigma_A{}^2 > \sigma^2$, indicating that H is false. Clearly, large values of F are indicative of the latter situation. Table I tells us how large F has to be to suggest rejection of H at a given significance level. Note that the appropriate column and row of Table I are determined by the number of degrees of freedom associated with the numerator and denominator of F.

The information required for performing an analysis of variance is most conveniently summarized in the *analysis of variance table* 19.3.

TABLE 19.3 **Analysis of Variance Table**

Source of Variation	Sum of Squares, SS	Degrees of Freedom, df	Mean Square, MS = SS/df
Among Samples	$SS_A = E - C$	$k - 1$	$MS_A = SS_A/(k-1)$
Within Samples	$SS_W = D - E$	$N - k$	$MS_W = SS_W/(N-k)$
Total	$SS_T = D - C$	$N - 1$	

We have discussed the normal theory analysis of data from a k-sample experiment of the simplest type, the completely randomized design in the terminology of Chapter 15. The analysis of more complicated data, for example, data from complete and incomplete block designs, is discussed in books dealing with advanced statistical methodology.

PROBLEMS

1 You have computed the following:

T_j:	454	549	425	351
n_j:	6	7	6	4

D = 139511

Find MS_A, MS_W, F. What is the critical value of F using $\alpha = .01$?

2 Compute F for the data in Problem 3 of Chapter 15. What is the critical value of F using $\alpha = .05$?

3 On the basis of Table 15.1, test the hypothesis that the mean scores for the three teaching methods are the same. (*Suggestion:* You can reduce computing effort by subtracting the same number, for example, 60, from each of the 21 observations. An even better choice is 75.)

4 Solve Problem 5 of Chapter 15, using normal theory methods.

5 Solve Problem 13 of Chapter 15 using normal theory methods.

6 Solve Problem 16 of Chapter 15 using normal theory methods.

20

REGRESSION AND CORRELATION

145 In Chapter 10 we dealt with the question of whether or not two characteristics used to classify sample items were related or not. In Chapter 16 we were concerned with measuring strength of relationship of two rankings. Now we assume that we have pairs of measurements involving two continuous variables, conventionally denoted by x and y. For instance, x may refer to a student's high school grade-point average, y, to the student's college grade-point average. Or x may be the height of a person and y, the weight. The Kendall and Spearman rank correlation coefficients of Chapter 16 can be used to measure the strength of relationship between two such variables. However, another correlation coefficient is more commonly used. We will define this correlation coefficient in Section 150. Now we are more interested in prediction: If we know the value of one variable, say x, what can we say about y? If x is a student's midterm grade, what does x reveal about the student's final course grade y?

Problems of prediction arise when an experimenter is interested in investigating changes in a variable y that result from changes in a variable x. If x measures the amount of fertilizer applied to a plot of land, how do crop yields y change with different amounts x of fertilizer? To answer this question, an experimenter will select n plots of land, apply predetermined amounts of fertilizer $x_1, x_2, \ldots, x_n$, not necessarily all different, and determine corresponding crop yields $y_1, y_2, \ldots, y_n$.

Quite generally we will assume that an investigator has selected n levels, $x_1, x_2, \ldots, x_n$ of a variable x and then measures corresponding values $y_1, y_2, \ldots, y_n$ of a variable y. When investigating the relationship of a variable y on a variable x on the basis of n observation pairs

$$(x_1, y_1), (x_2, y_2), \ldots, (x_n, y_n),$$

it is helpful to represent the n "data points" graphically by means of a *scatter diagram* in which the x-value of a given pair is plotted along a horizontal axis and the y-value along a vertical axis as in Figure 20.1. Figure 20.1 is based on $n = 13$ pairs of observations involving four different levels of the variable x.

FIGURE 20.1 Scatter Diagram

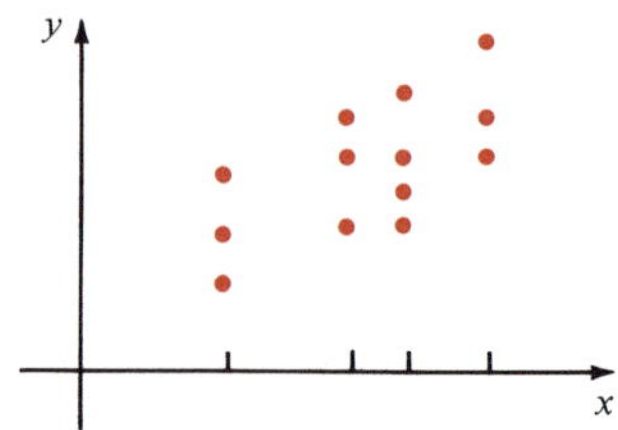

Linear Regression

146 Experience shows that many scatter diagrams exhibit a nearly linear trend. We therefore consider the following linear regression model: For every possible x-value, we imagine a population of y-values. If we denote the mean of the y-population corresponding to a given value x by $\mu_{y.x}$, then $\mu_{y.x}$ considered as a function of x represents a straight line.† We write

20.2 $$\mu_{y.x} = \alpha + \beta x,$$

where α and β are constants (not to be confused with type 1 and type 2 error probabilities). The quantities α and β in (20.2) have the following statistical meaning: α is the mean of the y-population when $x = 0$; β equals the amount of increase (decrease if β is negative) in $\mu_{y.x}$ brought about by an increase of one unit in the value of x. We call (20.2) the *regression of y on x*.

Our first problem is to find point estimates of the regression coefficients α and β and consequently of $\mu_{y.x}$.

147 *Least Square Estimates*

In the past when determining point estimates of population parameters, we usually appealed to intuition and common sense. When it comes to the estimation of the regression parameters α and β , something more than intuition is required. What we need is a basic principle or method

†The subscript $y.x$ indicates that we are dealing with the mean of a y-population, where the y-population may depend on the value of some measurement x.

that tells us how to go about finding appropriate estimates. For the type of problem with which we are concerned, the principle of least squares has been used by mathematicians for well over 100 years.

To illustrate the principle of least squares, we return briefly to the simpler problem of estimating the mean μ of a population on the basis of a random sample $y_1, \ldots, y_n$. In Section 131 we proposed to estimate the population mean μ as the average of sample observations, $\bar{y} = (y_1 + \cdots + y_n)/n$. Let m stand for the *least squares estimate* of μ. The principle of least squares states that the sum of squared deviations of the sample observations be smaller when taken from m than from any other number. More formally, the least squares estimate m minimizes the following sum of squares:

$$Q = \Sigma(y - m)^2.$$

In general, problems of minimization require the tools of calculus. But in the present case a noncalculus argument allows us to show that the least squares principle is satisfied for $m = \bar{y}$, so that the estimate (17.1) actually *is* the least squares estimate.

To prove our statement, we write $y - m = (y - \bar{y}) + (\bar{y} - m)$. Then

$$\begin{aligned} Q &= \Sigma[(y - \bar{y}) + (\bar{y} - m)]^2 \\ &= \Sigma[(y - \bar{y})^2 + 2(y - \bar{y})(\bar{y} - m) + (\bar{y} - m)^2] \\ &= \Sigma(y - \bar{y})^2 + 2(\bar{y} - m)\,\Sigma(y - \bar{y}) + n(\bar{y} - m)^2 \\ &= \Sigma(y - \bar{y})^2 + n(\bar{y} - m)^2, \end{aligned}$$

since $\Sigma(y - \bar{y}) = 0$. The last expression for Q is minimized when m is set equal to $\bar{y}$.

Point Estimates for α, β, and $\mu_{y.x}$

We will denote the point estimates of α, β, and $\mu_{y.x}$ by a, b, and $m_{y.x}$, respectively. In view of (20.2) the three estimates are related as

20.3 $$m_{y.x} = a + bx.$$

For our regression problem, the sum of squared deviations to be minimized according to the principle of least squares is

$$Q = \Sigma(y - m_{y.x})^2 = \Sigma(y - a - bx)^2.$$

By applying calculus methods (see Problem 17), it can be shown that the following expressions minimize Q:

20.4 $$b = \frac{\Sigma(x - \bar{x})(y - \bar{y})}{\Sigma(x - \bar{x})^2}$$

20.5 and $$a = \overline{y} - b\overline{x},$$

where b is given by (20.4). Substitution in (20.3) gives

20.6 $$m_{y.x} = \overline{y} + b(x - \overline{x}).$$

These are the least squares estimates of β, α, and $\mu_{y.x}$.

Let us introduce the following notation:

$$T_{xx} = \Sigma(x - \overline{x})^2 = \Sigma x^2 - \frac{1}{n}(\Sigma x)^2,$$

$$T_{xy} = \Sigma(x - \overline{x})(y - \overline{y}) = \Sigma xy - \overline{x}\,\Sigma y - \overline{y}\,\Sigma x + n\overline{x}\overline{y}$$

$$= \Sigma xy - \frac{1}{n}(\Sigma x)(\Sigma y),$$

and, for future use,

$$T_{yy} = \Sigma(y - \overline{y})^2 = \Sigma y^2 - \frac{1}{n}(\Sigma y)^2.$$

Then by (20.4), we have

20.7 $$b = \frac{T_{xy}}{T_{xx}}.$$

148 An Example

To investigate how well grades on the midterm examination of a statistics course predict grades on the final examination, a professor selects 27 students according to their midterm grades and finds that the distribution of final examination grades is as shown in Table 20.8.

TABLE 20.8

Midtern Grade (x)	45	55	65	75	85	95
Number of Students	1	4	6	6	6	4
Final Examination Grade (y)	52	54	57	62	89	94
		63	72	77	93	97
		60	77	91	97	83
		62	80	71	74	95
			61	89	80	
			75	70	66	

We determine that

$$n = 27$$
$$\Sigma x = 45 + 4 \times 55 + \cdot \cdot \cdot + 4 \times 95 = 1995$$
$$\overline{x} = 73.9$$
$$\Sigma x^2 = 45^2 + 4 \times 55^2 + \cdot \cdot \cdot + 4 \times 95^2 = 152675$$
$$\Sigma y = 52 + 54 + \cdot \cdot \cdot + 95 = 2041$$

$$\overline{y} = 75.6$$
$$\Sigma y^2 = 52^2 + 54^2 + \cdot \cdot \cdot + 95^2 = 159427$$
$$\Sigma xy = 45 \times 52 + 55 \times 54 + \cdot \cdot \cdot + 95 \times 95 = 154885,$$

and
$$T_{xx} = \Sigma x^2 - \frac{(\Sigma x)^2}{n} = 5266.7$$

$$T_{xy} = \Sigma xy - \frac{(\Sigma x)(\Sigma y)}{n} = 4077.8$$

$$T_{yy} = \Sigma y^2 - \frac{(\Sigma y)^2}{n} = 5142.5.$$

Thus
$$b = \frac{T_{xy}}{T_{xx}} = \frac{4077.8}{5266.7} = 0.774$$

and
$$\begin{aligned} m_{y.x} &= \overline{y} + b(x - \overline{x}) \\ &= 75.6 + 0.774(x - 73.9) \\ &= 18.4 + 0.774x \\ &= a + bx. \end{aligned}$$

This result can be used for prediction purposes. For students who have received 68 on the midterm examination, we predict a final examination grade of

$$18.4 + 0.774 \times 68 = 71.0.$$

For 88 on the midterm examination, we predict

$$18.4 + 0.774 \times 88 = 86.5$$

on the final examination.

Figure 20.9 shows the scatter diagram for the data in Table 20.8. The *least squares regression line*

$$m_{y.x} = 18.4 + 0.774x$$

has been drawn in to show the kind of fit the line represents. The regression estimates corresponding to $x = 68$ and $x = 88$ are also shown.

149 *The Index of Determination*

Figure 20.9 shows that we can get some idea of how well the regression line $m_{y.x} = \overline{y} + b(x - \overline{x})$ fits the given set of observations by superimposing the regression line on the scatter diagram. An objective measure of the closeness of fit is provided by the index of determination, which takes values between 0 and 1 and is denoted by r^2. The two limits $r^2 = 0$ and $r^2 = 1$ represent "no" fit and "perfect" fit, respectively. The exact meaning of these two terms will emerge as we go along.

As we look at Table 20.8, we note that there is considerable variation among final grades, which extend from a low of 52 to a high of 97. Part

FIGURE 20.9

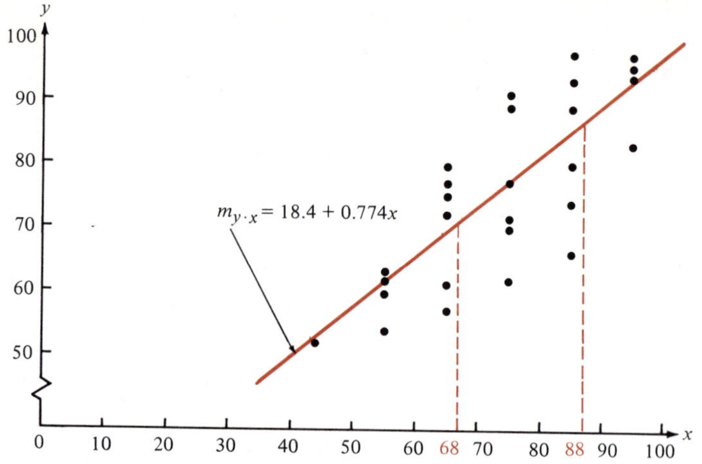

of this variation is caused by random fluctuation. But part of this variation is "explained" through the investigator's choice of students according to midterm grades. In general, we expect students with higher midterm grades to have higher final grades.

We divide the *total variation* among y-values as measured by $\Sigma(y - \bar{y})^2$ into *explained* and *unexplained variation.* Explained variation arises from the investigator's choice of x-values at which the variable y is observed. Such variation measures the strength of the relationship between x and y. Unexplained variation arises from random factors over which the investigator has no control, and manifests itself in that students who have identical midterm grades, in general, have different final examination grades.

We define the index of determination r^2 as the proportion that explained variation bears to total variation. To find a formal expression for r^2, we write for each pair (x_j, y_j), $j = 1, \ldots, n$,

$$y_j - \bar{y} = (y_j - m_{y.x_j}) + (m_{y.x_j} - \bar{y});$$

FIGURE 20.10

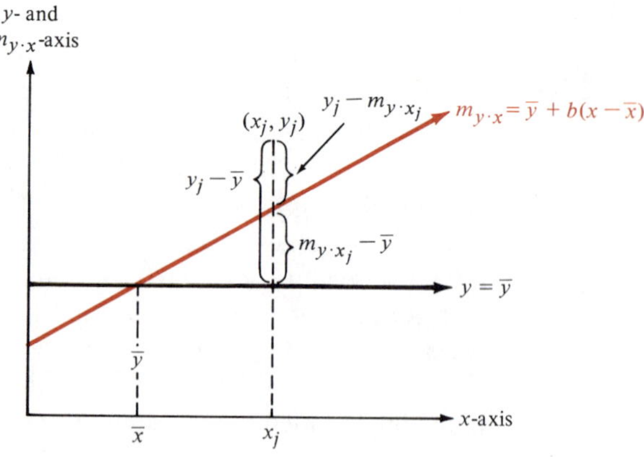

that is, we express the amount by which the observation y_j deviates from the mean of all y-observations as the sum of two components as shown in Figure 20.10. The first component measures by how much y_j deviates from the regression estimate $m_{y.x}$ at $x = x_j$, while the second component measures by how much this regression estimate deviates from $\bar{y}$. By (20.6) we have $m_{y.x_j} - \bar{y} = b(x_j - \bar{x})$, so that

$$y_j - \bar{y} = (y_j - m_{y.x_j}) + b(x_j - \bar{x}).$$

Squaring both sides and summing over j, we get

20.11 $$\Sigma(y - \bar{y})^2 = \Sigma(y - m_{y.x})^2 + b^2\Sigma(x - \bar{x})^2.$$

[The cross-product terms vanish on summation as follows:

$$\begin{aligned}\Sigma(y - m_{y.x})(x - \bar{x}) &= \Sigma[(y - \bar{y}) - b(x - \bar{x})](x - \bar{x}) \\ &= \Sigma(y - \bar{y})(x - \bar{x}) - b\Sigma(x - \bar{x})^2 = 0\end{aligned}$$

in view of (20.4).] In (20.11) the sum of squares $b^2\Sigma(x - \bar{x})^2$ arising from the component $m_{y.x} - \bar{y}$ represents *explained* variation or variation *due to regression*, as it is sometimes called. The sum of squares $\Sigma(y - m_{y.x})^2$ represents *unexplained* variation or variation *about regression*, since it measures deviation from the line of regression. We then have

20.12 $$\begin{aligned}r^2 &= \frac{b^2\Sigma(x - \bar{x})^2}{\Sigma(y - \bar{y})^2} \\ &= \frac{(T_{xy}/T_{xx})^2 T_{xx}}{T_{yy}} = \frac{T_{xy}{}^2}{T_{xx}T_{yy}}.\end{aligned}$$

For our numerical example, we obtain

$$r^2 = \frac{4077.8^2}{5266.7 \times 5142.5} = .61,$$

so that roughly 60 percent of the variation among final examination scores is caused by variation in midterm grades, while the remaining 40 percent represents random fluctuation.

According to (20.12), $r^2 = 1$ means that $\Sigma(y - \bar{y})^2 = b^2\Sigma(x - \bar{x})^2$, but then by (20.11),

$$\Sigma(y - m_{y.x})^2 = 0.$$

This means that, for every observation pair (x_j, y_j), we have

$$y_j - m_{y.x_j} = 0.$$

All points in the scatter diagram lie along the regression line as in Figure 20.13. This we call perfect fit.

We have $r^2 = 0$ if $b = 0$. But then by (20.6), we obtain

$$m_{y.x} = \bar{y} = \text{constant}$$

FIGURE 20.13

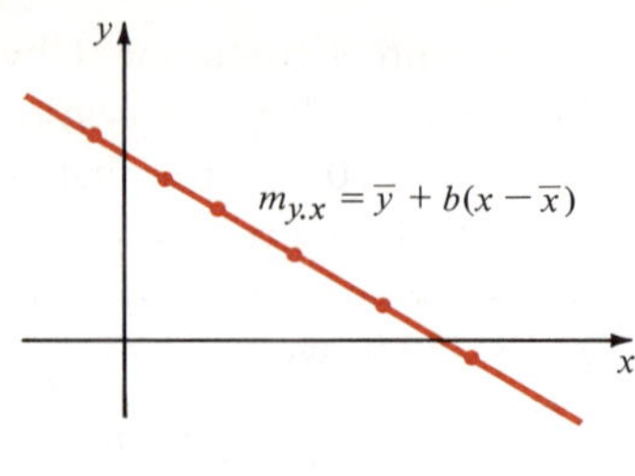

and $$\Sigma(y - m_{y.x})^2 = \Sigma(y - \bar{y})^2.$$

The regression line is horizontal and none of the total variation can be explained in terms of regression of y on x. Whatever the value x at which we observe y, our prediction is always $\bar{y}$. Knowledge of x does not help in predicting y. This is the meaning of what we called "no" fit.

The Correlation Coefficient r

150 The square root of the index of determination,

20.14
$$r = b\sqrt{\frac{\Sigma(x - \bar{x})^2}{\Sigma(y - \bar{y})^2}} = \frac{\Sigma(x - \bar{x})(y - \bar{y})}{\sqrt{\Sigma(x - \bar{x})^2\Sigma(y - \bar{y})^2}} = \frac{T_{xy}}{\sqrt{T_{xx}T_{yy}}}$$

is called the Pearson product moment correlation coefficient, or simply the correlation coefficient. It is the most widely used (and abused) of all correlation coefficients. Our discussion of the index of determination shows that r, or rather r^2, measures reduction in sampling variation due to *linear* regression. If the assumption of linearity of regression is unwarranted, use of r as a measure of strength of relationship between x and y may be very misleading.

Since r^2 varies between 0 and 1, r varies between -1 and $+1$. According to (20.14), r and b have the same sign: When b is positive, so is r; when b is negative, so is r; when b equals zero, so does r. As a consequence, a positive r implies that x and y tend to move in the same direction; a negative value of r implies that x and y tend to move in opposite directions.

The extremes, -1 and $+1$, indicate that all sample points (x_j, y_j) lie on the line of regression. Thus Figure 20.13 represents the case $r = -1$. The Kendall and Spearman rank correlation coefficients take the value $+1$ for identical rankings of x- and y-observations. They take the value -1 for reverse rankings. Since $r = +1$ implies identical rankings and $r = -1$ implies reverse rankings, the two rank correlation coefficients take extreme values whenever r does. The reverse, however, is not true. It is possible for the two rank correlation coefficients to take the value $+1$ (or -1) and for r to be smaller than $+1$ (or greater than -1); see Problem 8.

A Test of Independence

151 If in (20.2), β equals 0,

$$\mu_{y.x} = \alpha = \text{constant}$$

and the mean of y-observations has the same value irrespective of the x-value that the experimenter has chosen. The experimenter may just as well ignore the variable x. We say that *y is independent of x.*

How can we test for independence? In Chapter 16 we based tests of independence on the Kendall and Spearman rank correlation coefficients. In the case of linear regression, a test can also be based on the regression coefficient b or the correlation coefficient r, provided that we make a normality assumption involving variables y.

Our only assumption so far has been the validity of (20.2). The principle of least squares led to the estimate (20.6) for $\mu_{y.x}$. To test for independence, we now make the additional assumption that the distribution of the y-variables is normal with (unknown) variance σ^2. A theorem of mathematical statistics then states:

THEOREM 20.15 The estimate b of β is normally distributed with mean β and variance $\sigma^2/\Sigma(x - \bar{x})^2$.

Using the approach discussed in Section 140, we can conclude that

20.16
$$t = \frac{b\sqrt{\Sigma(x - \bar{x})^2}}{s}$$

has a t-distribution, where s^2 is an appropriate estimate of σ^2. When $y_1, \ldots, y_n$ is a random sample from a population with mean μ and variance σ^2, we estimate σ^2 as

$$s^2 = \frac{1}{n-1}\Sigma(y - \bar{y})^2,$$

where $\bar{y}$ is the least squares estimate of the population mean μ. By analogy, in our regression model, we estimate the variance σ^2 as

20.17
$$s^2 = \frac{1}{n-2}\Sigma(y - m_{y.x})^2,$$

where $m_{y.x}$ is the least squares estimate (20.6) of $\mu_{y.x}$. According to (20.11), we have

$$\Sigma(y - m_{y.x})^2 = \Sigma(y - \overline{y})^2 - b^2\Sigma(x - \overline{x})^2$$
$$= T_{yy} - \frac{T_{xy}^2}{T_{xx}}.$$

The appropriate number of degrees of freedom associated with the estimate (20.17), and therefore with the t-statistic (20.16), is $n - 2$.

For the numerical example of Section 148, we find that

$$s^2 = \left(T_{yy} - \frac{T_{xy}^2}{T_{xx}}\right)/(n - 2)$$
$$= \left(5142.5 - \frac{4077.8^2}{5266.7}\right)/25$$
$$= 79.41;$$

and
$$t = \frac{0.774\sqrt{5266.7}}{\sqrt{79.41}} = 6.3$$

For 25 degrees of freedom, such a large value of t implies rejection of the hypothesis of independence of y on x.

The t-statistic (20.16) can be expressed in terms of the correlation coefficient r. Indeed,

$$t^2 = \frac{b^2\Sigma(x - \overline{x})^2}{s^2} = \frac{(n - 2)b^2\Sigma(x - \overline{x})^2}{\Sigma(y - m_{y.x})^2} = \frac{(n - 2)r^2}{1 - r^2}$$

so that

20.18
$$t = \frac{r\sqrt{n - 2}}{\sqrt{1 - r^2}}.$$

Numerically large values of t, which suggest rejection of the hypothesis of independence, correspond to numerically large values of r. It is possible to compute r_γ-values corresponding to t_γ-values such that the hypothesis of independence is rejected at significance level $1 - \gamma$ whenever $r < -r_\gamma$ or $r > r_\gamma$.

Chance Fluctuations in x and y

152 So far we have assumed that the experimenter selects n values $x_1, \ldots, x_n$ and then observes values $y_1, \ldots, y_n$ corresponding to these x-values. Actually, in practice, the experimental setup may be quite different. Thus if x refers to a student's verbal score on the GRE (Graduate Record Examination) and y, to the student's quantitative score, we may select n students at random from the population of students taking the GRE and determine the n pairs of observations $(x_1, y_1), \ldots, (x_n, y_n)$. In such an experiment the x-values are not predetermined but are left to chance.

Statisticians then may speak of a correlation model rather than a regression model—a distinction that is primarily historical and has little to recommend itself.

The earlier regression analysis remains valid even when x-values are not preselected but are determined by chance. However, in the latter case the correlation coefficient r also serves as an estimate of a parameter ρ, called the population correlation coefficient. This coefficient characterizes how much better one can predict y from x compared to not knowing x or, vice versa, how much better one can predict x from y compared to not knowing y. When analyzing data where both x's and y's are subject to chance, investigators often tend to concentrate primarily on ρ rather than on predicting one variable from the other.

When both x- and y-variables are subject to chance, not only can we predict y from x, but also x from y. In the Graduate Record Examination example, there is no reason to limit ourselves to an investigation of how quantitative scores change with verbal scores. It is just as reasonable to ask how verbal scores change with quantitative scores.

We then deal with two different regression lines. To avoid confusion, it is helpful to adopt a precise notation. When talking of the regression of y on x, we write

$$\mu_{y.x} = \alpha_{y.x} + \beta_{y.x}x,$$

which is estimated as

$$m_{y.x} = \bar{y} + b_{y.x}(x - \bar{x}),$$

where $b_{y.x}$ (our former b) is given by (20.4). Correspondingly, the regression of x on y is given by

$$\mu_{x.y} = \alpha_{x.y} + \beta_{x.y}y,$$

which is estimated as

$$m_{x.y} = \bar{x} + b_{x.y}(y - \bar{y}),$$

where $$b_{x.y} = \frac{T_{xy}}{T_{yy}} = \frac{\Sigma(x - \bar{x})(y - \bar{y})}{\Sigma(y - \bar{y})^2}.$$

As far as the correlation coefficient r is concerned, (20.14) shows that r is symmetric in x and y and therefore that r remains unchanged under an interchange of x and y. Under the present setup, the test based on (20.18) becomes a test of the hypothesis $\rho = 0$, which implies that it does not pay to try predicting y from x or x from y.

General Regression Models

153* The regression problem discussed in Section 146 considers the simplest possible regression setup, where the mean of a variable y is a linear function of another variable x. We may try to generalize this situation by considering the regression of y on k variables $x_1, \ldots, x_k$. If y is the college grade-point average of a student, x_1 may be the student's high school grade-point average; x_2, the verbal Scholastic Aptitude Test score; x_3, the quantitative Scholastic Aptitude Test score. Clearly, other variables can be added, like IQ scores, class standing in high school, and so on. We then consider the following regression model for the mean of y-observations:

20.19 $$\mu_{y.x_1 \ldots x_k} = \beta_0 + \beta_1 x_1 + \cdots + \beta_k x_k,$$

where $\beta_0, \beta_1, \ldots, \beta_k$ are unknown regression coefficients. The problem is to find estimates for, and test hypotheses about, these parameters.

While model (20.19) is linear in $x_1, \ldots, x_k$, it actually covers a much wider range. Thus if we set $x_1 = x$, $x_2 = x^2, \ldots, x_k = x^k$, we have the case of polynomial regression of y on a single variable x, where we assume that

20.20 $$\mu_{y.x} = \beta_0 + \beta_1 x + \beta_2 x^2 + \cdots + \beta_k x^k.$$

By combining models (20.19) and (20.20), we can consider polynomial regression in more than one x-variable. The analysis of such problems is discussed in books dealing with advanced statistical methodology.

PROBLEMS

1 For the four number pairs

(1,1) (2,3) (3,6) (4,10),

compute

a. Σx, Σy;
b. Σx^2, Σxy, Σy^2;
c. T_{xx}, T_{xy}, T_{yy};
d. $T_{xy}^2/T_{xx}T_{yy}$.

2 For the following number pairs

(1,6) (2,5) (3,3) (4,3) (6,1),

find

a. $m_{y.x}$,
b. $m_{y.5}$,
c. r^2,
d. r, and
e. test for independence

3 Assume that you have computed the following information for 16 pairs of observations:

$$\Sigma x = 896 \qquad \Sigma y = 655$$
$$\Sigma x^2 = 52330 \qquad \Sigma y^2 = 29652 \qquad \Sigma xy = 38368.$$

Answer the following questions, making any appropriate assumptions.

a. Find the regression of y on x.
b. Estimate σ^2.
c. Estimate $\mu_{y.50}$.
d. Compute r.
e. What proportion of the total variability of y-observations can be explained in terms of the regression of y on x?
f. Test for independence.

4 Find the correlation coefficient r for the data of Problem 5 of Chapter 16.

5 Use the data of Problem 6 of Chapter 16

a. to find the regression of diastolic blood pressure on systolic blood pressure;
b. to predict the diastolic blood pressure of a person whose systolic blood pressure equals 110;
c. to find the correlation between systolic and diastolic blood pressure.

6 For the data in Table 16.2,

a. find the regression of psychology grades on statistics grades;
b. find the regression of statistics grades on psychology grades;
c. superimpose the two regression lines on the scatter diagram;
d. test for independence between psychology and statistics grades; what assumptions does your test imply?

7 For the data of Problem 13 of Chapter 16,

a. find the Pearson product moment correlation coefficient;
b. test for independence.

8 For the four numbers pairs of Problem 1,

a. find r;
b. find the Kendall rank correlation coefficient;
c. find the Spearman rank correlation coefficient.
d. What do these results show?

9 Analyze the height–weight data you were asked to collect for Problem 7 of Chapter 12 by the methods of this chapter.

10 Find r for the data of Problem 16 of Chapter 16.

11 Using the data of Problem 18 of Chapter 16,

a. find the regression of nicotine content on tar;
b. test for independence.

12 In Problem 10 of Chapter 16, use the ranks of typist 1 as x-observations and the ranks of typist 2 as y-observations. Compute the correlation coefficient r. Compare your result with the Spearman rank correlation coefficient for the same data. (See Problem 20a of Chapter 16.)

13 Repeat the previous problem using the data of Problem 11 of Chapter 16. What general result do you suspect?

14 For a sample of size 11, find r_γ such that we reject the hypothesis of independence at significance level $1 - \gamma = .01$ whenever $|r| > r_\gamma$.

15 Consider a scatter diagram with the least square regression line superimposed. What is the specific meaning of the quantity $Q = \Sigma(y - m_{y.x})^2$?

Problems 16 and 17 are for students with some knowledge of calculus.

16 Use calculus methods to show that $Q = \Sigma(y - m)^2$, where summation extends over $y_1, \ldots, y_n$, is a minimum when $m = \bar{y}$.

17 Let $Q = \Sigma(y - a - bx)^2$ where summation extends over n number pairs (x_j, y_j).

a. Show that

$$\frac{\partial Q}{\partial a} = -2\Sigma(y - a - bx)$$

$$\frac{\partial Q}{\partial b} = -2\Sigma x(y - a - bx).$$

b. Show that the system of equations

$$\frac{\partial Q}{\partial a} = 0, \qquad \frac{\partial Q}{\partial b} = 0$$

has solutions b and a given by (20.4) and (20.5), respectively.

c. Why does it follow that Q is a minimum for these values of a and b?

APPENDIX

Summation Notation

154 In statistical work it is frequently necessary to compute sums of numbers. It is often convenient to use "mathematical shorthand" to indicate summation. Suppose that we have m numbers

$$u_1, u_2, \ldots, u_m,$$

which we may also indicate as

$$u_i, \qquad i = 1, 2, \ldots, m.$$

We will write

$$\Sigma u$$

(read: "summation u") in place of the sum

$$u_1 + u_2 + \cdots + u_m.$$

A few additional examples will illustrate the summation notation and at the same time indicate some basic properties of summation. Let a and b be two arbitrary constants. Then

$$\begin{aligned}\Sigma(a + bu) &= (a + bu_1) + (a + bu_2) + \cdots + (a + bu_m) \\ &= (a + a + \cdots + a) + b(u_1 + u_2 + \cdots + u_m) \\ &= ma + b\Sigma u.\end{aligned}$$

If

$$v_i, \qquad i = 1, 2, \ldots, m$$

is a second set of m numbers, then

$$\Sigma uv = u_1v_1 + u_2v_2 + \cdots + u_mv_m$$

and

$$\begin{aligned}\Sigma(u - v) &= (u_1 - v_1) + (u_2 - v_2) + \cdots + (u_m - v_m)\\ &= (u_1 + u_2 + \cdots + u_m) - (v_1 + v_2 + \cdots + v_m)\\ &= \Sigma u - \Sigma v.\end{aligned}$$

In some statistical applications it is necessary to consider numbers indexed by two subscripts i and j such as

$$u_{ij}, \qquad i = 1, 2, \ldots, r; \quad j = 1, 2, \ldots, c.$$

For example, if we have c different sets of numbers and each set contains r numbers, u_{ij} may stand for the ith number in the jth set. We usually want to be able to indicate summation over individual sets of numbers and therefore require a more elaborate summation notation. Thus

$$\Sigma_i u_{ij} = u_{1j} + u_{2j} + \cdots + u_{rj}$$

and

$$\Sigma_j u_{ij} = u_{i1} + u_{i2} + \cdots + u_{ic}.$$

To denote summation over all $r \times c$ terms, we write

$$\Sigma_i \Sigma_j u_{ij} = u_{11} + u_{12} + \cdots + u_{rc}.$$

Thus if we have numbers

$$a_1, a_2, \ldots, a_r$$

and

$$b_1, b_2, \ldots, b_c$$

and want the sum of all possible cross products $u_{ij} = a_i b_j$, we write

$$\begin{aligned}\Sigma_i \Sigma_j u_{ij} &= \Sigma_i \Sigma_j a_i b_j\\ &= a_1(b_1 + b_2 + \cdots + b_c) + a_2(b_1 + b_2 + \cdots + b_c)\\ &\qquad + \cdots \qquad + a_r(b_1 + b_2 + \cdots + b_c)\\ &= (a_1 + a_2 + \cdots + a_r)(b_1 + b_2 + \cdots + b_c)\\ &= (\Sigma a)(\Sigma b).\end{aligned}$$

Events and Simple Events

155 When discussing probability in this book, we use the term "event" in the usual colloquial sense and presumably few, if any, readers stop to wonder what is meant by an event. For mathematical purposes such an intuitive approach is unsatisfactory. We must say what we mean by *event* before we can talk rigorously of the probability of an event.

In probability we are concerned with chance experiments, experiments whose exact outcome cannot be predicted with certainty. Thus in Chapter 3 the experiment usually consists in the selection of one student from among a group of students. While we cannot predict which student will be selected, we can give a precise listing of all possibilities. Quite generally, we assume that all possibilities, or *outcomes,* associated with a chance experiment are known, and we use the symbols $e_1, e_2, \ldots$ to denote the possible outcomes of the experiment under discussion. Often the number of possible outcomes is finite. In such a case we use the symbol t to denote the total number of possible outcomes. On the other hand, there are experiments for which it is necessary or convenient to consider infinitely many outcomes. In such a case it is physically impossible to list all outcomes.

We can now define what we mean by event. Any collection of outcomes associated with a chance experiment is called an event. In particular, a single outcome determines an event, often called a *simple event.* An *arbitrary event* can be considered as an aggregate of simple events.

A rigorous treatment of probability is relatively simple if there are only a finite number of outcomes. The case of an infinite number of outcomes, particularly the noncountable case, presents additional mathematical problems, some of which are mentioned briefly in Chapter 11. To keep our treatment simple, we assume throughout the remainder of this discussion that the number of possible outcomes is finite. With the necessary mathematical preparation, relevant concepts can be redefined so that they apply in the infinite case.

Consider a random experiment with outcomes (or simple events) $e_1, \ldots, e_t$. Associated with each simple event e_j, where $j = 1, \ldots, t$, is a nonnegative number p_j called the probability of the simple event e_j, in symbols, $p_j = P(e_j)$, such that

$$p_1 + \cdots + p_t = 1.$$

If A is an arbitrary event, we set

$$P(A) = \sum_{e_j \text{ in } A} p_j,$$

where the summation extends only over the simple events e_j that make up the event A.

Random Variables

156 In practice often we are not interested in the particular outcome e_j of our experiment, but only in some numerical feature associated with e_j. Thus in the taxi example of Chapter 1, the e_j are the actual taxis in town. But we are interested only in one specific feature of the taxi, the number on its shield. Such features as color, make, or size are of no concern to us, at least for the particular problem under consideration.

An association that assigns a unique numerical value to every outcome of a chance experiment defines a *random* (or *chance*) *variable.* It is customary to use capital letters X, Y, Z, etc. to denote random variables. However, it is helpful to remember that, when we write X, we really mean $X = x(e)$, a function whose argument e is the outcome of the chance experiment under consideration.

Associated with a random variable X is a function $f(x)$ called the distribution of X. For a given value x, $f(x)$ is the probability with which the random variable X takes the specific value x: $f(x) = P(X = x)$. Since we are considering only experiments with a finite number of outcomes, a random variable X can take only a finite number of different values. Let us denote these values by $x_1, \ldots, x_m$. Then for $i = 1, \ldots, m$, we have

A.1
$$f(x_i) = P(X = x_i) = \sum_{x(e_j) = x_i} p_j,$$

the sum of the probabilities of the simple events e_j, where the random variable X takes the value x_i. Since X takes one and only one value for each simple event e_j, we have

$$f(x_1) + \cdots + f(x_m) = p_1 + \cdots + p_t = 1.$$

EXAMPLE A.2 *Rolls of Two Fair Dice* Our experiment consists of rolling two fair dice. For definiteness, suppose that one is red and the other green. The simple events of this experiment are the 36 number pairs (1,1), (1,2), . . . , (r,g), . . . , (6,6), where the first number, r, in each pair indicates the number of points appearing on the red die and the second number, g, indicates the number of points appearing on the green die. The dice rolls are said to be *fair* if each of these 36 simple events has the same probability, namely $1/36$. For many dice games, the only quantity of interest is the number of points on both dice. If we denote this number by X, then X is a random variable defined for the simple events $e = (r,g)$, such that $X = x(e) = r + g$. The possible values of X are 2,3, . . . , 12. To find the distribution of X, we need to know $f(x) = P(X = x)$ for $x = 2,3, \ldots, 12$. For fair dice, simple enumeration gives the distribution

A.3

x	2	3	4	5	6	7	8	9	10	11	12
$f(x)$	$1/36$	$2/36$	$3/36$	$4/36$	$5/36$	$6/36$	$5/36$	$4/36$	$3/36$	$2/36$	$1/36$

EXAMPLE A.4 Consider trials with only two possible outcomes called success and failure, respectively. Let K equal the number of successes in n independent trials. Then K is a random variable with possible values $k = 0,1, \ldots, n$. In Chapter 4 we derive the distribution of the random variable K. [In Chapter 4 we write $b(k) = P(K = k)$, rather than $f(k)$.]

EXAMPLE A.5 *Random Variables and Statistical Inference* We can illustrate how the concept of random variable enters into statistical inference by looking once more at the taxi problem. Consider the experiment that consists in the random selection of one taxi from among the t taxis in town. The possible outcomes $e_1, \ldots, e_t$ of this experiment are the various taxis in town. We can then define a random variable W such that $W = w(e)$ equals the number on the taxi shield. The distribution of W is given by $f(w) = P(W = w) = 1/t$ for $w = 1,2, \ldots, t$. In the taxi problem, the number t is unknown. In Chapters 1 and 3 we discuss how to estimate the parameter t of the taxi problem as well as test hypotheses about t using five observed values for the random variable W.

157 *The Expected Value of a Random Variable*

A random variable X is characterized by its distribution function. However, instead of knowing the complete distribution of X, it is often sufficient and even desirable to have only some summary information about the distribution. We usually want to know some typical or central value and perhaps to have some indication by how much individual values of X are likely to deviate from this typical value. There is no unique way of defining the center of the distribution of a random variable. In Chapter 12 we use the median as our center. The median divides the population into two equal parts on the basis of probability. But other measures of centrality can be defined. Corresponding to the center of gravity of a physical system is the expected value of a random variable.

The expected value, or *mean*, of a random variable X, denoted variously by the symbols μ, μ_x, or $E(X)$ (read, the expected value of X), is defined as the weighted sum

A.6
$$E(X) = \mu = x_1P(X = x_1) + \cdots + x_mP(X = x_m) \\ = x_1f(x_1) + \cdots + x_mf(x_m) = \Sigma xf(x),$$

where again m is the number of different values the random variable X can take.

EXAMPLE A.2 *cont'd.* For the sum of points of two fair dice, we find from (A.3),

$$\mu = 2 \times \frac{1}{36} + 3 \times \frac{2}{36} + \cdots + 12 \times \frac{1}{36} = 7.$$

The significance of μ as a typical or central value is seen when we consider what happens when we repeat the chance experiment producing a value of X a large number of times and average the resulting values. Let $\#(x_i)$, $i = 1, \ldots, m$, stand for the number of times we observe the value x_i in n repetitions of the experiment. The average of all observed x-values (the term "average" understood in the sense of the arithmetic mean) is

$$\text{Average} = \frac{x_1 \#(x_1) + \cdots + x_m \#(x_m)}{n}$$

$$= x_1 \frac{\#(x_1)}{n} + \cdots + x_m \frac{\#(x_m)}{n}$$

$$\doteq x_1 f(x_1) + \cdots + x_m f(x_m) = \mu,$$

according to the frequency interpretation of probability. Exactly in the same way that $f(x_i)$ may be looked upon as the long-run relative frequency of the value x_i, the mean μ may be looked upon as the long-run average value of X. If X is the amount of money a gambler wins or loses in a game of chance, μ is the average amount of money the gambler can *expect* to win or lose per game in a large number of games. The term *expected value* goes back to the time when the primary application of probability was to games of chance.

While the mean μ is a typical value of X in the sense described above, it should be noted that μ need not be equal to any of the possible values of X. For rolls of a single fair die, $\mu = 3.5$.

The concept of expected value can be extended. If $h(x)$ is a function of the argument x and X is a random variable, then $h(X)$ is a random variable and its expected value is given by the weighted sum

$$E[h(X)] = \Sigma h(x) f(x)$$

where as before $f(x) = P(X = x)$.

A particularly simple and useful result is obtained when $h(x)$ is a linear function, that is, $h(x) = a + bx$, where a and b are constants. Then

$$\begin{aligned} E(a + bX) &= \Sigma(a + bx)f(x) \\ &= \Sigma af(x) + \Sigma bxf(x) \\ &= a\Sigma f(x) + b\Sigma xf(x) = a + bE(X). \end{aligned}$$

We have proved the following important result:

THEOREM A.7 The expected value of a linear function of a random variable equals the linear function of the expected value of the random variable.

158 *The Variance*

There is no unique way of characterizing variability of a random variable X about its mean μ. One very useful measure of variability is the variance denoted by σ^2, $\sigma_X{}^2$, or $V(X)$. The variance of a random variable X is defined as the expected value of $(X - \mu)^2$,

$$\sigma^2 = E(X - \mu)^2 = \Sigma(x - \mu)^2 f(x).$$

EXAMPLE A.2 *cont'd.* For the sum of points of two fair dice, we find that

$$\sigma^2 = (2-7)^2 \times \frac{1}{36} + (3-7)^2 \times \frac{2}{36} + \cdots + (12-7)^2 \times \frac{1}{36} = \frac{35}{6}.$$

By an analysis similar to the one involving the mean, we note that the variance may be considered as an average of squared deviations from the mean μ in infinitely many experiments. It follows that the variance measures variability about the mean. A small value of σ^2 indicates that large deviations from the mean are very unlikely to occur, while a large value of σ^2 indicates that large deviations from the mean are not only possible, but also likely, to occur.

For practical applications, it is often more useful to measure variability in terms of a quantity expressed in the same dimensionality as the actual observations (rather than the square of the observations). Thus for practical purposes, the positive square root of the variance, the standard deviation denoted by σ, is often used as a measure of variability about the mean.

The definition of σ^2 given above is primarily useful for theoretical purposes. For purposes of computation, another form is often preferable. With the help of Theorem A.8 we find that

$$\begin{aligned} \sigma^2 = E(X-\mu)^2 &= E(X^2 - 2\mu X + \mu^2) \\ &= E(X^2) - 2\mu E(X) + \mu^2 \\ &= E(X^2) - 2\mu^2 + \mu^2 \\ &= E(X^2) - \mu^2. \end{aligned}$$

EXAMPLE A.2 *cont'd.* Recomputation for the dice example gives

$$\sigma^2 = \left(2^2 \times \frac{1}{36} + 3^2 \times \frac{2}{36} + \cdots + 12^2 \times \frac{1}{36}\right) - 7^2 = \frac{35}{6}$$

as before.

Expected Value and Variance of Sums of Two Random Variables

159 If we substitute (A.1) in (A.6), we find that the expected value of a random variable X can be computed as

$$E(X) = x(e_1)p_1 + \cdots + x(e_t)p_t.$$

Similarly, if X and Y are two random variables, we have

$$\begin{aligned} E(X+Y) &= [x(e_1)+y(e_1)]p_1 + \cdots + [x(e_t)+y(e_t)]p_t \\ &= [x(e_1)p_1 + \cdots + x(e_t)p_t] \\ &\quad + [y(e_1)p_1 + \cdots + y(e_t)p_t] \\ &= E(X) + E(Y). \end{aligned}$$

We have proved the following important result:

THEOREM A.8 The expected value of the sum of two random variables equals the sum of the expected values of the two random variables.

EXAMPLE A.2 *cont'd.* The random variable X of Example A.2 can be written as $X = R + G$, where R equals the number of points on the red die and G equals the number of points on the green die. Then by Theorem A.8,

$$E(X) = E(R) + E(G) = \frac{7}{2} + \frac{7}{2} = 7,$$

as before.

160 We now turn to the expectation of the product of two random variables X and Y. Problem 12 shows that it is not generally true that the expected value of the product of two random variables equals the product of the individual expected values. However, the statement is true for independent random variables:

DEFINITION A.9 Two random variables X and Y with values x_i, $i = 1, \ldots, m$, and y_j, $j = 1, \ldots, n$, respectively, are said to be *independent* if for all possible combinations (x_i, y_j) we have

$$P(X = x_i \text{ and } Y = y_j) = P(X = x_i)P(Y = y_j). \tag{A.10}$$

THEOREM A.11 If X and Y are two independent random variables, then the expected value of the product XY equals the product of the expected values,

$$E(XY) = E(X)E(Y).$$

Indeed, using (A.6) and (A.10), we have

$$\begin{aligned} E(XY) &= \Sigma_i\Sigma_j x_i y_j P(X = x_i \text{ and } Y = y_j) \\ &= \Sigma_i\Sigma_j x_i y_j P(X = x_i)P(Y = y_j) \\ &= [\Sigma_i x_i P(X = x_i)][\Sigma_j y_j P(Y = y_j)] \\ &= E(X)E(Y). \end{aligned}$$

Theorem A.11 allows us to prove the following important result:

THEOREM A.12 If X and Y are two independent random variables, then the variance of the sum $X + Y$ equals the sum of the individual variances,

$$V(X + Y) = V(X) + V(Y).$$

(Note that Theorem A.8 is true whether random variables X and Y are independent or not.) By the definition of variance and Theorem A.8, we have

$$\begin{aligned} V(X+Y) &= E[(X+Y)-\mu_{x+y}]^2 \\ &= E[(X+Y)-(\mu_x+\mu_y)]^2 \\ &= E[(X-\mu_x)+(Y-\mu_y)]^2 \\ &= E(X-\mu_x)^2 + E(Y-\mu_y)^2 + 2E[(X-\mu_x)(Y-\mu_y)] \\ &= V(X)+V(Y)+2C(X,Y), \end{aligned}$$

where $C(X, Y) = E[(X-\mu_x)(Y-\mu_y)]$ is the *covariance* of X and Y. We have to show that the covariance vanishes for independent random variables. This result follows immediately when we apply Theorem A.11 to the result of Problem 13.

EXAMPLE A.2 *cont'd.* Earlier we computed the variance of the variable X defined in Example A.2 by using the distribution (A.3) of the random variable X. Since X can be written as $X = R + G$ and R and G are independent random variables, Theorem A.12 gives

$$V(X) = V(R) + V(G) = \frac{35}{12} + \frac{35}{12} = \frac{35}{6}$$

with much less computation.

161 Theorems A.8 and A.12 immediately extend to sums of more than two random variables.

EXAMPLE A.13 Consider the random variable K of Example A.4. K can be represented as

$$K = I_1 + \cdot\cdot\cdot + I_n,$$

where I_j equals 1 or 0 depending on whether the jth trial ends in success or failure, $j = 1, \ldots, n$. According to Problem 11, we have $E(I_j) = p$ and $V(I_j) = pq$. Then

$$\begin{aligned} E(K) &= E(I_1) + \cdot\cdot\cdot + E(I_n) \\ &= p \quad + \cdot\cdot\cdot + p = np \end{aligned}$$

and

$$\begin{aligned} V(K) &= V(I_1) + \cdot\cdot\cdot + V(I_n) \\ &= pq \quad + \cdot\cdot\cdot + pq = npq. \end{aligned}$$

The same results can be obtained much more laboriously using the binomial distribution (4.1).

PROBLEMS

1 Consider the random variable K of Example A.4. What are the experimental outcomes e that determine the value of K? How many such outcomes are there? (*Hint:* See Chapter 4).

For Problems 2 through 7 use the following information. Let R equal the number of points on a red die, G, those on a green die. Always assume that dice tosses are fair.

2 Write out the distribution of R.

3 Show that $E(R) = 7/2$.

4 Show that $E(R - 7/2) = 0$.

5 Show that $V(R) = 35/12$.

6 If, in Problems 2 through 5, R is replaced by G, what are the corresponding answers? Why?

7 Find the distribution of $X = R + G$.

8 If X is any random variable, show that $E(X - \mu) = 0$. It follows that $E(X - \mu)$ is useless as a measure of variability of X about its mean.

9 Let c be a constant. Show that $E(X - c)^2$ is a minimum when $c = E(X) = \mu$. (*Hint:* Write $E(X - c)^2 = E[(X - \mu) + (\mu - c)]^2$.)

10 If the random variable X has variance σ_x^2 and if a and b are constants, show that the random variable $Y = a + bX$ has variance $\sigma_y^2 = b^2\sigma_x^2$.

11 Let I be a random variable that takes only the values 1 and 0 with probabilities p and $q = 1 - p$, $0 < p < 1$. (Such a variable is often called an *indicator variable.*) Show that

a. $E(I) = E(I^2) = p$,
b. $V(I) = pq$.

12 The random variable U takes the values $+1$ and -1 with probability $1/2$ each.

a. Find $E(U)$.
b. Find $E(U^2)$.
c. Use the results of parts a and b to show that

$$E(U^2) = E(U \times U) \neq E(U) \times E(U),$$

so that in this case the expected value of a product does not equal the product of the expected values.

13 Let X and Y be two random variables. Show that we have the following computing formula for the covariance of X and Y:

$$C(X, Y) = E[(X - \mu_x)(Y - \mu_y)] = E(XY) - \mu_x\mu_y.$$

TABLES

ACKNOWLEDGMENTS

The author is grateful for receiving permission to reproduce or adapt the following tables from the indicated sources:

Tables D, H, and I from *Biometrika Tables for Statisticians,* Volume 1 (ed. 3), Cambridge University Press, 1967.

Tables F and G from F. Wilcoxon, S. Katti, and R. A. Wilcox, Critical Values and Probability Levels for the Wilcoxon Rank Sum Test and the Wilcoxon Signed Rank Test. Reprinted with permission of the publisher, The American Mathematical Society, from *Selected Tables in Mathematical Statistics.* Copyright © 1973, Volume 1, pp. 129–169 and 237–259.

Table J from *A Million Random Digits with 100,000 Normal Deviates,* The RAND Corporation, 1955.

TABLE A **Binomial Probabilities: $n = 10$**

k \ p	.05	.10	.20	.30	⅓	.40	.50	
0	.599	.349	.107	.028	.017	.006	.001	10
1	.315	.387	.268	.121	.087	.040	.010	9
2	.075	.194	.302	.233	.195	.121	.044	8
3	.010	.057	.201	.267	.260	.215	.117	7
4	.001	.011	.088	.200	.228	.251	.205	6
5		.001	.026	.103	.137	.201	.246	5
6			.006	.037	.057	.111	.205	4
7			.001	.009	.016	.042	.117	3
8				.001	.003	.011	.044	2
9						.002	.010	1
10							.001	0
	.95	.90	.80	.70	⅔	.60	.50	p \ k

Binomial Probabilities: $n = 13$

k \ p	.05	.10	.20	.30	⅓	.40	.50	
0	.513	.254	.055	.010	.005	.001		13
1	.351	.367	.179	.054	.034	.011	.002	12
2	.111	.245	.268	.139	.100	.045	.010	11
3	.021	.100	.246	.218	.183	.111	.035	10
4	.003	.028	.154	.234	.230	.184	.087	9
5		.006	.069	.180	.207	.221	.157	8
6		.001	.023	.103	.137	.197	.209	7
7			.006	.044	.069	.131	.209	6
8			.001	.014	.026	.066	.157	5
9				.003	.007	.024	.087	4
10				.001	.002	.006	.035	3
11						.001	.010	2
12							.002	1
13								0
	.95	.90	.80	.70	⅔	.60	.50	p \ k

Table A cont'd. **Binomial Probabilities: $n = 20$**

k \ p	.05	.10	.20	.30	⅓	.40	.50	
0	.358	.122	.012	.001				20
1	.377	.270	.058	.007	.003			19
2	.189	.285	.137	.028	.014	.003		18
3	.060	.190	.205	.072	.043	.012	.001	17
4	.013	.090	.218	.130	.091	.035	.005	16
5	.002	.032	.175	.179	.146	.075	.015	15
6		.009	.109	.192	.182	.124	.037	14
7		.002	.055	.164	.182	.166	.074	13
8			.022	.114	.148	.180	.120	12
9			.007	.065	.097	.160	.160	11
10			.002	.031	.054	.117	.176	10
11				.012	.025	.071	.160	9
12				.004	.009	.035	.120	8
13				.001	.003	.015	.074	7
14					.001	.005	.037	6
15						.001	.015	5
16							.005	4
17							.001	3
	.95	.90	.80	.70	⅔	.60	.50	p \ k

Binomial Probabilities: $n = 25$

k \ p	.05	.10	.20	.30	⅓	.40	.50	
0	.277	.072	.004					25
1	.365	.199	.024	.001				24
2	.231	.266	.071	.007	.003			23
3	.093	.226	.136	.024	.011	.002		22
4	.027	.138	.187	.057	.031	.007		21
5	.006	.065	.196	.103	.066	.020	.002	20
6	.001	.024	.163	.147	.110	.044	.005	19
7		.007	.111	.171	.149	.080	.014	18
8		,002	.062	.165	.167	.120	.032	17
9			.029	.134	.158	.151	.061	16
10			.012	.092	.126	.161	.097	15
11			.004	.054	.086	.147	.133	14
12			.001	.027	.050	.114	.155	13
13				.011	.025	.076	.155	12
14				.004	.011	.043	.133	11
15				.001	.004	.021	.097	10
16					.001	.009	.061	9
17						.003	.032	8
18						.001	.014	7
19							.005	6
20							.002	5
	.95	.90	.80	.70	⅔	.60	.50	p \ k

TABLE B **Normal Curve Areas**

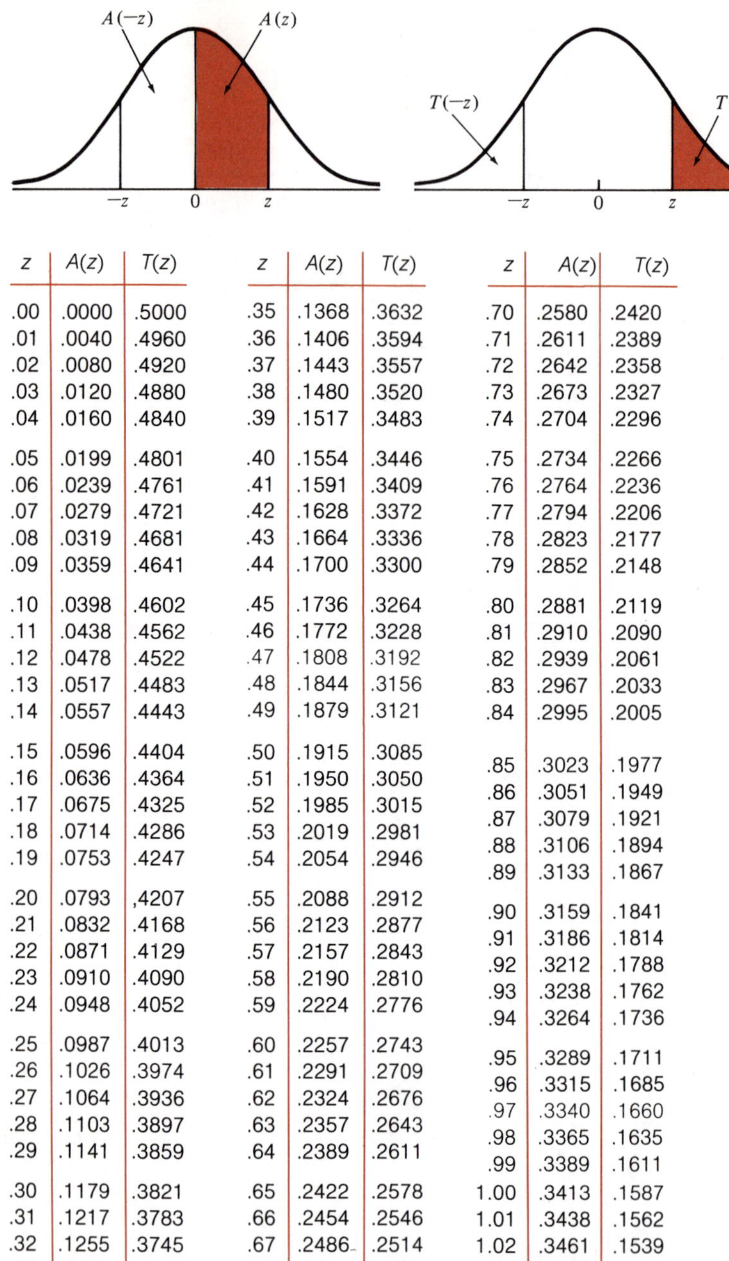

z	A(z)	T(z)	z	A(z)	T(z)	z	A(z)	T(z)
.00	.0000	.5000	.35	.1368	.3632	.70	.2580	.2420
.01	.0040	.4960	.36	.1406	.3594	.71	.2611	.2389
.02	.0080	.4920	.37	.1443	.3557	.72	.2642	.2358
.03	.0120	.4880	.38	.1480	.3520	.73	.2673	.2327
.04	.0160	.4840	.39	.1517	.3483	.74	.2704	.2296
.05	.0199	.4801	.40	.1554	.3446	.75	.2734	.2266
.06	.0239	.4761	.41	.1591	.3409	.76	.2764	.2236
.07	.0279	.4721	.42	.1628	.3372	.77	.2794	.2206
.08	.0319	.4681	.43	.1664	.3336	.78	.2823	.2177
.09	.0359	.4641	.44	.1700	.3300	.79	.2852	.2148
.10	.0398	.4602	.45	.1736	.3264	.80	.2881	.2119
.11	.0438	.4562	.46	.1772	.3228	.81	.2910	.2090
.12	.0478	.4522	.47	.1808	.3192	.82	.2939	.2061
.13	.0517	.4483	.48	.1844	.3156	.83	.2967	.2033
.14	.0557	.4443	.49	.1879	.3121	.84	.2995	.2005
.15	.0596	.4404	.50	.1915	.3085	.85	.3023	.1977
.16	.0636	.4364	.51	.1950	.3050	.86	.3051	.1949
.17	.0675	.4325	.52	.1985	.3015	.87	.3079	.1921
.18	.0714	.4286	.53	.2019	.2981	.88	.3106	.1894
.19	.0753	.4247	.54	.2054	.2946	.89	.3133	.1867
.20	.0793	,4207	.55	.2088	.2912	.90	.3159	.1841
.21	.0832	.4168	.56	.2123	.2877	.91	.3186	.1814
.22	.0871	.4129	.57	.2157	.2843	.92	.3212	.1788
.23	.0910	.4090	.58	.2190	.2810	.93	.3238	.1762
.24	.0948	.4052	.59	.2224	.2776	.94	.3264	.1736
.25	.0987	.4013	.60	.2257	.2743	.95	.3289	.1711
.26	.1026	.3974	.61	.2291	.2709	.96	.3315	.1685
.27	.1064	.3936	.62	.2324	.2676	.97	.3340	.1660
.28	.1103	.3897	.63	.2357	.2643	.98	.3365	.1635
.29	.1141	.3859	.64	.2389	.2611	.99	.3389	.1611
.30	.1179	.3821	.65	.2422	.2578	1.00	.3413	.1587
.31	.1217	.3783	.66	.2454	.2546	1.01	.3438	.1562
.32	.1255	.3745	.67	.2486	.2514	1.02	.3461	.1539
.33	.1293	.3707	.68	.2517	.2483	1.03	.3485	.1515
.34	.1331	.3669	.69	.2549	.2451	1.04	.3508	.1492

z	$A(z)$	$T(z)$	z	$A(z)$	$T(z)$	z	$A(z)$	$T(z)$
1.05	.3531	.1469	1.40	.4192	.0808	1.75	.4599	.0401
1.06	.3554	.1446	1.41	.4207	.0793	1.76	.4608	.0392
1.07	.3577	.1423	1.42	.4222	.0778	1.77	.4616	.0384
1.08	.3599	.1401	1.43	.4236	.0764	1.78	.4625	.0375
1.09	.3621	.1379	1.44	.4251	.0749	1.79	.4633	.0367
1.10	.3643	.1357	1.45	.4265	.0735	1.80	.4641	.0359
1.11	.3665	.1335	1.46	.4279	.0721	1.81	.4649	.0351
1.12	.3686	.1314	1.47	.4292	.0708	1.82	.4656	.0344
1.13	.3708	.1292	1.48	.4306	.0694	1.83	.4664	.0336
1.14	.3729	.1271	1.49	.4319	.0681	1.84	.4671	.0329
1.15	.3749	.1251	1.50	.4332	.0668	1.85	.4678	.0322
1.16	.3770	.1230	1.51	.4345	.0655	1.86	.4686	.0314
1.17	.3790	.1210	1.52	.4357	.0643	1.87	.4693	.0307
1.18	.3810	.1190	1.53	.4370	.0630	1.88	.4699	.0301
1.19	.3830	.1170	1.54	.4382	.0618	1.89	.4706	.0294
1.20	.3849	.1151	1.55	.4394	.0606	1.90	.4713	.0287
1.21	.3869	.1131	1.56	.4406	.0594	1.91	.4719	.0281
1.22	.3888	.1112	1.57	.4418	.0582	1.92	.4726	.0274
1.23	.3907	.1093	1.58	.4429	.0571	1.93	.4732	.0268
1.24	.3925	.1075	1.59	.4441	.0559	1.94	.4738	.0262
1.25	.3944	.1056	1.60	.4452	.0548	1.95	.4744	.0256
1.26	.3962	.1038	1.61	.4463	.0537	1.96	.4750	.0250
1.27	.3980	.1020	1.62	.4474	.0526	1.97	.4756	.0244
1.28	.3997	.1003	1.63	.4484	.0516	1.98	.4761	.0239
1.29	.4015	.0985	1.64	.4495	.0505	1.99	.4767	.0233
1.30	.4032	.0968	1.65	.4505	.0495	2.00	.4772	.0228
1.31	.4049	.0951	1.66	.4515	.0485	2.01	.4778	.0222
1.32	.4066	.0934	1.67	.4525	.0475	2.02	.4783	.0217
1.33	.4082	.0918	1.68	.4535	.0465	2.03	.4788	.0212
1.34	.4099	.0901	1.69	.4545	.0455	2.04	.4793	.0207
1.35	.4115	.0885	1.70	.4554	.0446	2.05	.4798	.0202
1.36	.4131	.0869	1.71	.4564	.0436	2.06	.4803	.0197
1.37	.4147	.0853	1.72	.4573	.0427	2.07	.4808	.0192
1.38	.4162	.0838	1.73	.4582	.0418	2.08	.4812	.0188
1.39	.4177	.0823	1.74	.4591	.0409	2.09	.4817	.0183

z	$A(z)$	$T(z)$	z	$A(z)$	$T(z)$	z	$A(z)$	$T(z)$
2.10	.4821	.0179	2.45	.4929	.0071	2.80	.4974	.0026
2.11	.4826	.0174	2.46	.4931	.0069	2.81	.4975	.0025
2.12	.4830	.0170	2.47	.4932	.0068	2.82	.4976	.0024
2.13	.4834	.0166	2.48	.4934	.0066	2.83	.4977	.0023
2.14	.4838	.0162	2.49	.4936	.0064	2.84	.4977	.0023
2.15	.4842	.0158	2.50	.4938	.0062	2.85	.4978	.0022
2.16	.4846	.0154	2.51	.4940	.0060	2.86	.4979	.0021
2.17	.4850	.0150	2.52	.4941	.0059	2.87	.4979	.0021
2.18	.4854	.0146	2.53	.4943	.0057	2.88	.4980	.0020
2.19	.4857	.0143	2.54	.4945	.0055	2.89	.4981	.0019
2.20	.4861	.0139	2.55	.4946	.0054	2.90	.4981	.0019
2.21	.4864	.0136	2.56	.4948	.0052	2.91	.4982	.0018
2.22	.4868	.0132	2.57	.4949	.0051	2.92	.4983	.0017
2.23	.4871	.0129	2.58	.4951	.0049	2.93	.4983	.0017
2.24	.4875	.0125	2.59	.4952	.0048	2.94	.4984	.0016
2.25	.4878	.0122	2.60	.4953	.0047	2.95	.4984	.0016
2.26	.4881	.0119	2.61	.4955	.0045	2.96	.4985	.0015
2.27	.4884	.0116	2.62	.4956	.0044	2.97	.4985	.0015
2.28	.4887	.0113	2.63	.4957	.0043	2.98	.4986	.0014
2.29	.4890	.0110	2.64	.4959	.0041	2.99	.4986	.0014
2.30	.4893	.0107	2.65	.4960	.0040	3.00	.4987	.0013
2.31	.4896	.0104	2.66	.4961	.0039	3.05	.4989	.0011
2.32	.4898	.0102	2.67	.4962	.0038	3.10	.4990	.0010
2.33	.4901	.0099	2.68	.4963	.0037	3.15	.4992	.0008
2.34	.4904	.0096	2.69	.4964	.0036	3.20	.4993	.0007
2.35	.4906	.0094	2.70	.4965	.0035	3.25	.4994	.0006
2.36	.4909	.0091	2.71	.4966	.0034	3.30	.4995	.0005
2.37	.4911	.0089	2.72	.4967	.0033	3.35	.4996	.0004
2.38	.4913	.0087	2.73	.4968	.0032	3.40	.4997	.0003
2.39	.4916	.0084	2.74	.4969	.0031	3.45	.4997	.0003
2.40	.4918	.0082	2.75	.4970	.0030	3.50	.4998	.0002
2.41	.4920	.0080	2.76	.4971	.0029	3.60	.4998	.0002
2.42	.4922	.0078	2.77	.4972	.0028	3.70	.4999	.0001
2.43	.4925	.0075	2.78	.4973	.0027	3.80	.4999	.0001
2.44	.4927	.0073	2.79	.4974	.0026	3.90	.5000	.0000

TABLE C **Normal Deviations**

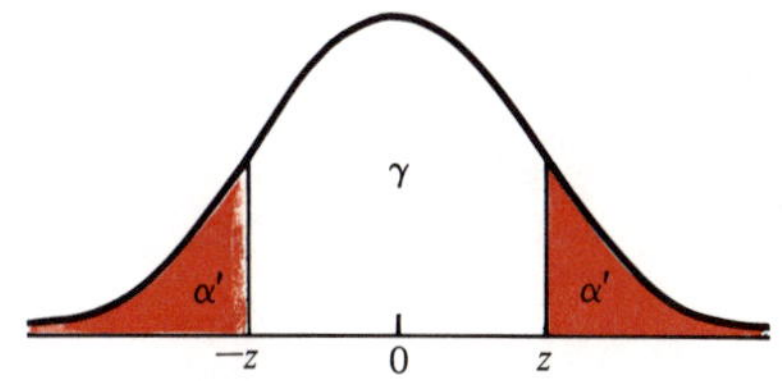

γ	α''	α'	z
.995	.005	.0025	2.807
.99	.01	.005	2.576
.985	.015	.0075	2.432
.98	.02	.01	2.326
.975	.025	.0125	2.241
.97	.03	.015	2.170
.965	.035	.0175	2.108
.96	.04	.02	2.054
.954	.046	.023	2.000
.95	.05	.025	1.960
.94	.06	.03	1.881
.92	.08	.04	1.751
.9	.1	.05	1.645
.85	.15	.075	1.440
.8	.2	.10	1.282
.75	.25	.125	1.150
.7	.3	.150	1.036
.6	.4	.20	0.842
.5	.5	.25	0.674
.4	.6	.30	0.524
.3	.7	.35	0.385
.2	.8	.40	0.253
.1	.9	.45	0.126

γ = area between $-z$ and z
= confidence coefficient

$\alpha' = \frac{1}{2}(1 - \gamma)$
= area above z
= area below $-z$
= significance level for one-sided test

$\alpha'' = 1 - \gamma = 2\alpha'$
= area beyond $-z$ and z
= significance level for two-sided test

TABLE D **Chi-square Distribution**

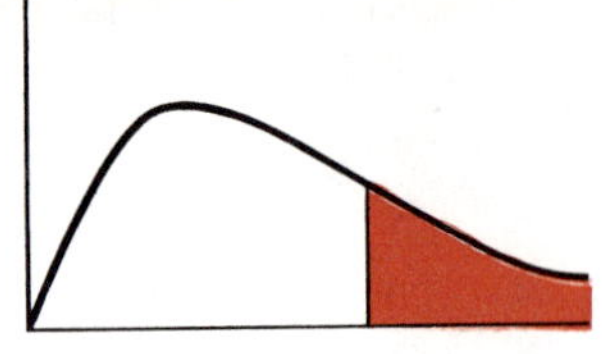

df	.990	.975	.950	.900	.750	.500	.250	.100	.050	.025	.010
	Upper Tail Probabilities										
1		.001	.004	.016	.102	.455	1.32	2.71	3.84	5.02	6.63
2	.020	.051	.103	.211	.575	1.39	2.77	4.61	5.99	7.38	9.21
3	.115	.216	.352	.584	1.21	2.37	4.11	6.25	7.81	9.35	11.3
4	.297	.484	.711	1.06	1.92	3.36	5.39	7.78	9.49	11.1	13.3
5	.554	.831	1.15	1.61	2.67	4.35	6.63	9.24	11.1	12.8	15.1
6	.872	1.24	1.64	2.20	3.45	5.35	7.84	10.6	12.6	14.4	16.8
7	1.24	1.69	2.17	2.83	4.25	6.35	9.04	12.0	14.1	16.0	18.5
8	1.65	2.18	2.73	3.49	5.07	7.34	10.2	13.4	15.5	17.5	20.1
9	2.09	2.70	3.33	4.17	5.90	8.34	11.4	14.7	16.9	19.0	21.7
10	2.56	3.25	3.94	4.87	6.74	9.34	12.5	16.0	18.3	20.5	23.2
11	3.05	3.82	4.57	5.58	7.58	10.3	13.7	17.3	19.7	21.9	24.7
12	3.57	4.40	5.23	6.30	8.44	11.3	14.8	18.5	21.0	23.3	26.2
13	4.11	5.01	5.89	7.04	9.30	12.3	16.0	19.8	22.4	24.7	27.7
14	4.66	5.63	6.57	7.79	10.2	13.3	17.1	21.1	23.7	26.1	29.1
15	5.23	6.26	7.26	8.55	11.0	14.3	18.2	22.3	25.0	27.5	30.6
16	5.81	6.91	7.96	9.31	11.9	15.3	19.4	23.5	26.3	28.8	32.0
17	6.41	7.56	8.67	10.1	12.8	16.3	20.5	24.8	27.6	30.2	33.4
18	7.01	8.23	9.39	10.9	13.7	17.3	21.6	26.0	28.9	31.5	34.8
19	7.63	8.91	10.1	11.7	14.6	18.3	22.7	27.2	30.1	32.9	36.2
20	8.26	9.59	10.9	12.4	15.5	19.3	23.8	28.4	31.4	34.2	37.6
21	8.90	10.3	11.6	13.2	16.3	20.3	24.9	29.6	32.7	35.5	38.9
22	9.54	11.0	12.3	14.0	17.2	21.3	26.0	30.8	33.9	36.8	40.3
23	10.2	11.7	13.1	14.8	18.1	22.3	27.1	32.0	35.2	38.1	41.6
24	10.9	12.4	13.8	15.7	19.0	23.3	28.2	33.2	36.4	39.4	43.0
25	11.5	13.1	14.6	16.5	19.9	24.3	29.3	34.4	37.7	40.6	44.3
26	12.2	13.8	15.4	17.3	20.8	25.3	30.4	35.6	38.9	41.9	45.6
27	12.9	14.6	16.2	18.1	21.7	26.3	31.5	36.7	40.1	43.2	47.0
28	13.6	15.3	16.9	18.9	22.7	27.3	32.6	37.9	41.3	44.5	48.3
29	14.3	16.0	17.7	19.8	23.6	28.3	33.7	39.1	42.6	45.7	49.6
30	15.0	16.8	18.5	20.6	24.5	29.3	34.8	40.3	43.8	47.0	50.9
40	22.2	24.4	26.5	29.1	33.7	39.3	45.6	51.8	55.8	59.3	63.7
60	37.5	40.5	43.2	46.5	52.3	59.3	67.0	74.4	79.1	83.3	88.4
	.010	.025	.050	.100	.250	.500	.750	.900	.950	.975	.990
	Lower Tail Probabilities										

For large *df*, the tabulated value equals approximately $\frac{1}{2}(z + \sqrt{2(df) - 1})^2$, where z is obtained from Table C corresponding to the appropriate upper or lower tail probability α'.

TABLE E *d*-Factors for Sign Test and Confidence Intervals for the Median

γ = confidence coefficient
$\alpha' = ½(1 - \gamma)$ = one-sided significance level
$\alpha'' = 2\alpha' = 1 - \gamma$ = two-sided significance level

n	d	γ	α''	α'
3	1	.750	.250	.125
4	1	.875	.125	.062
5	1	.938	.062	.031
6	1	.969	.031	.016
	2	.781	.219	.109
7	1	.984	.016	.008
	2	.875	.125	.063
8	1	.992	.008	.004
	2	.930	.070	.035
	3	.711	.289	.145
9	1	.996	.004	.002
	2	.961	.039	.020
	3	.820	.180	.090
10	1	.998	.002	.001
	2	.979	.021	.011
	3	.891	.109	.055
11	1	.999	.001	.000
	2	.988	.012	.006
	3	.935	.065	.033
	4	.773	.227	.113
12	2	.994	.006	.003
	3	.961	.039	.019
	4	.854	.146	.073
13	2	.997	.003	.002
	3	.978	.022	.011
	4	.908	.092	.046
	5	.733	.267	.133
14	2	.998	.002	.001
	3	.987	.013	.006
	4	.943	.057	.029
	5	.820	.180	.090
15	3	.993	.007	.004
	4	.965	.035	.018
	5	.882	.118	.059
16	3	.996	.004	.002
	4	.979	.021	.011
	5	.923	.077	.038
	6	.790	.210	.105

n	d	γ	α''	α'
17	3	.998	.002	.001
	4	.987	.013	.006
	5	.951	.049	.025
	6	.857	.143	.072
18	4	.992	.008	.004
	5	.969	.031	.015
	6	.904	.096	.048
	7	.762	.238	.119
19	4	.996	.004	.002
	5	.981	.019	.010
	6	.936	.064	.032
	7	.833	.167	.084
20	4	.997	.003	.001
	5	.988	.012	.006
	6	.959	.041	.021
	7	.885	.115	.058
21	5	.993	.007	.004
	6	.973	.027	.013
	7	.922	.078	.039
	8	.811	.189	.095
22	5	.996	.004	.002
	6	.983	.017	.008
	7	.948	.052	.026
	8	.866	.134	.067
23	5	.997	.003	.001
	6	.989	.011	.005
	7	.965	.035	.017
	8	.907	.093	.047
	9	.790	.210	.105
24	6	.993	.007	.003
	7	.977	.023	.011
	8	.936	.064	.032
	9	.848	.152	.076
25	6	.996	.004	.002
	7	.985	.015	.007
	8	.957	.043	.022
	9	.892	.108	.054
26	7	.991	.009	.005
	8	.971	.029	.014
	9	.924	.076	.038
	10	.831	.169	.084

Table E cont'd.

n	d	γ	α''	α'
27	7	.994	.006	.003
	8	.981	.019	.010
	9	.948	.052	.026
	10	.878	.122	.061
28	7	.996	.004	.002
	8	.987	.013	.006
	9	.964	.036	.018
	10	.913	.087	.044
	11	.815	.185	.092
29	8	.992	.008	.004
	9	.976	.024	.012
	10	.939	.061	.031
	11	.864	.136	.068
30	8	.995	.005	.003
	9	.984	.016	.008
	10	.957	.043	.021
	11	.901	.099	.049
	12	.800	.200	.100
31	8	.997	.003	.002
	9	.989	.011	.005
	10	.971	.029	.015
	11	.929	.071	.035
	12	.850	.150	.075
32	9	.993	.007	.004
	10	.980	.020	.010
	11	.950	.050	.025
	12	.890	.110	.055
33	9	.995	.005	.002
	10	.986	.014	.007
	11	.965	.035	.018
	12	.920	.080	.040
	13	.837	.163	.081
34	10	.991	.009	.005
	11	.976	.024	.012
	12	.942	.058	.029
	13	.879	.121	.061
35	10	.994	.006	.003
	11	.983	.017	.008
	12	.959	.041	.020
	13	.910	.090	.045
	14	.825	.175	.088

n	d	γ	α''	α'
36	10	.996	.004	.002
	11	.989	.011	.006
	12	.971	.029	.014
	13	.935	.065	.033
	14	.868	.132	.066
37	11	.992	.008	.004
	12	.980	.020	.010
	13	.953	.047	.024
	14	.901	.099	.049
	15	.812	.188	.094
38	11	.995	.005	.003
	12	.986	.014	.007
	13	.966	.034	.017
	14	.927	.073	.036
	15	.857	.143	.072
39	12	.991	.009	.005
	13	.976	.024	.012
	14	.947	.053	.027
	15	.892	.108	.054
40	12	.994	.006	.003
	13	.983	.017	.008
	14	.962	.038	.019
	15	.919	.081	.040
	16	.846	.154	.077
41	12	.996	.004	.002
	13	.988	.012	.006
	14	.972	.028	.014
	15	.940	.060	.030
	16	.883	.117	.059
42	13	.992	.008	.004
	14	.980	.020	.010
	15	.956	.044	.022
	16	.912	.088	.044
	17	.836	.164	.082
43	13	.995	.005	.003
	14	.986	.014	.007
	15	.968	.032	.016
	16	.934	.066	.033
	17	.874	.126	.063
44	14	.990	.010	.005
	15	.977	.023	.011
	16	.951	.049	.024
	17	.904	.096	.048
	18	.826	.174	.087

n	d	γ	α''	α'	n	d	γ	α''	α'
45	14	.993	.007	.003	48	15	.994	.006	.003
	15	.984	.016	.008		16	.987	.013	.007
	16	.964	.036	.018		17	.971	.029	.015
	17	.928	.072	.036		18	.941	.059	.030
	18	.865	.135	.068		19	.889	.111	.056
46	14	.995	.005	.002	49	16	.991	.009	.005
	15	.989	.011	.006		17	.979	.021	.011
	16	.974	.026	.013		18	.956	.044	.022
	17	.946	.054	.027		19	.915	.085	.043
	18	.896	.104	.052		20	.848	.152	.076
47	15	.992	.008	.004	50	16	.993	.007	.003
	16	.981	.019	.009		17	.985	.015	.008
	17	.960	.040	.020		18	.967	.033	.016
	18	.921	.079	.039		19	.935	.065	.032
	19	.856	.144	.072		20	.881	.119	.059

For $n > 50$ use $d \doteq \frac{1}{2}(n + 1 - z\sqrt{n})$, where z is read from Table C.

TABLE F *d*-Factors for Wilcoxon Signed Rank Test and Confidence Intervals for the Median

γ = confidence coefficient
$\alpha' = \frac{1}{2}(1 - \gamma)$ = one-sided significance level
$\alpha'' = 2\alpha' = 1 - \gamma$ = two-sided significance level

n	d	γ	α''	α'	n	d	γ	α''	α'
3	1	.750	.250	.125	12	8	.991	.009	.005
4	1	.875	.125	.062		9	.988	.012	.006
5	1	.938	.062	.031		14	.958	.042	.021
	2	.875	.125	.063		15	.948	.052	.026
6	1	.969	.031	.016		18	.908	.092	.046
	2	.937	.063	.031		19	.890	.110	.055
	3	.906	.094	.047	13	10	.992	.008	.004
	4	.844	.156	.078		11	.990	.010	.005
7	1	.984	.016	.008		18	.952	.048	.024
	3	.953	.047	.016		19	.943	.057	.029
	4	.922	.078	.039		22	.906	.094	.047
	5	.891	.109	.055		23	.890	.110	.055
8	1	.992	.008	.004	14	13	.991	.009	.004
	2	.984	.016	.008		14	.989	.011	.005
	4	.961	.039	.020		22	.951	.049	.025
	5	.945	.055	.027		23	.942	.058	.029
	6	.922	.078	.039		26	.909	.091	.045
	7	.891	.109	.055		27	.896	.104	.052
9	2	.992	.008	.004	15	16	.992	.008	.004
	3	.988	.012	.006		17	.990	.010	.005
	6	.961	.039	.020		26	.952	.048	.024
	7	.945	.055	.027		27	.945	.055	.028
	9	.902	.098	.049		31	.905	.095	.047
	10	.871	.129	.065		32	.893	.107	.054
10	4	.990	.010	.005	16	20	.991	.009	.005
	5	.986	.014	.007		21	.989	.011	.006
	9	.951	.049	.024		30	.956	.044	.022
	10	.936	.064	.032		31	.949	.051	.025
	11	.916	.084	.042		36	.907	.093	.047
	12	.895	.105	.053		37	.895	.105	.052
11	6	.990	.010	.005	17	24	.991	.009	.005
	7	.986	.014	.007		25	.989	.011	.006
	11	.958	.042	.021		35	.955	.045	.022
	12	.946	.054	.027		36	.949	.051	.025
	14	.917	.083	.042		42	.902	.098	.049
	15	.898	.102	.051		43	.891	.109	.054

n	d	γ	α''	α'	n	d	γ	α''	α'
18	28	.991	.009	.005	22	49	.991	.009	.005
	29	.990	.010	.005		50	.990	.010	.005
	41	.952	.048	.024		66	.954	.046	.023
	42	.946	.054	.027		67	.950	.050	.025
	48	.901	.099	.049		76	.902	.098	.049
	49	.892	.108	.054		77	.895	.105	.053
19	33	.991	.009	.005	23	55	.991	.009	.005
	34	.989	.011	.005		56	.990	.010	.005
	47	.951	.049	.025		74	.952	.048	.024
	48	.945	.055	.027		75	.948	.052	.026
	54	.904	.096	.048		84	.902	.098	.049
	55	.896	.104	.052		85	.895	.105	.052
20	38	.991	.009	.005	24	62	.990	.010	.005
	39	.989	.011	.005		63	.989	.011	.005
	53	.952	.048	.024		82	.951	.049	.025
	54	.947	.053	.027		83	.947	.053	.026
	61	.903	.097	.049		92	.905	.095	.048
	62	.895	.105	.053		93	.899	.101	.051
21	43	.991	.009	.005	25	69	.990	.010	.005
	44	.990	.010	.005		70	.989	.011	.005
	59	.954	.046	.023		90	.952	.048	.024
	60	.950	.050	.025		91	.948	.052	.026
	68	.904	.096	.048		101	.904	.096	.048
	69	.897	.103	.052		102	.899	.101	.051

For $n > 25$ use

$$d \doteq \tfrac{1}{2}[\tfrac{1}{2}n(n+1) + 1 - z\sqrt{n(n+1)(2n+1)/6}],$$

where z is read from Table C.

TABLE G d-Factors for Wilcoxon-Mann-Whitney Test and Confidence Intervals for the Shift Parameter Δ

γ = confidence coefficient

$\alpha' = \frac{1}{2}(1 - \gamma)$ = one-sided significance level

$\alpha'' = 2\alpha' = 1 - \gamma$ = two-sided significance level

For sample sizes m and n beyond the range of this table use $d \doteq \frac{1}{2}\,[mn + 1 - z\sqrt{mn(m + n + 1)/3}]$, where z is read from Table C.

Size of Larger Sample	Size of Smaller Sample 3				4			
	d	γ	α''	α'	d	γ	α''	α'
3	1	.900	.100	.050				
4	1	.943	.057	.029	1	.971	.029	.014
	2	.886	.114	.057	2	.943	.057	.029
					3	.886	.114	.057
5	1	.964	.036	.018	1	.984	.016	.008
	2	.929	.071	.036	2	.968	.032	.016
	3	.857	.143	.071	3	.937	.063	.032
					4	.889	.111	.056
6	1	.976	.024	.012	1	.990	.010	.005
	2	.952	.048	.024	2	.981	.019	.010
	3	.905	.095	.048	3	.962	.038	.019
	4	.833	.167	.083	4	.933	.067	.033
					5	.886	.114	.057
7	1	.983	.017	.008	1	.994	.006	.003
	2	.967	.033	.017	2	.988	.012	.006
	3	.933	.067	.033	4	.958	.042	.021
	4	.883	.117	.058	5	.927	.073	.036
					6	.891	.109	.055
8	1	.988	.012	.006	2	.992	.008	.004
	3	.952	.048	.024	3	.984	.016	.008
	4	.915	.085	.042	5	.952	.048	.024
	5	.867	.133	.067	6	.927	.073	.036
					7	.891	.109	.055
9	1	.991	.009	.005	2	.994	.006	.003
	2	.982	.018	.009	3	.989	.011	.006
	3	.964	.036	.018	5	.966	.034	.017
	4	.936	.064	.032	6	.950	.050	.025
	5	.900	.100	.050	7	.924	.076	.038
					8	.894	.106	.053
10	1	.993	.007	.004	3	.992	.008	.004
	2	.986	.014	.007	4	.986	.014	.007
	4	.951	.049	.025	6	.964	.036	.018
	5	.923	.077	.039	7	.946	.054	.027
	6	.888	.112	.056	8	.924	.076	.038
					9	.894	.106	.053
11	1	.995	.005	.003	3	.994	.006	.003
	2	.989	.011	.006	4	.990	.010	.005
	4	.962	.038	.019	7	.960	.040	.020
	5	.940	.060	.030	8	.944	.056	.028
	6	.912	.088	.044	9	.922	.078	.039
	7	.874	.126	.063	10	.896	.104	.052
12	2	.991	.009	.004	4	.992	.008	.004
	3	.982	.018	.009	5	.987	.013	.007
	5	.952	.048	.024	8	.958	.042	.021
	6	.930	.070	.035	9	.942	.058	.029
	7	.899	.101	.051	10	.922	.078	.039
					11	.897	.103	.052

Size of Larger Sample	Size of Smaller Sample															
	5				6				7				8			
	d	γ	α''	α'	d	γ	α''	α'	d	γ	α''	α'	d	γ	α''	α'
5	1	.992	.008	.004												
	2	.984	.016	.008												
	3	.968	.032	.016												
	4	.994	.056	.028												
	5	.905	.095	.048												
	6	.849	.151	.075												
6	2	.991	.009	.004	3	.991	.009	.004								
	3	.983	.017	.009	4	.985	.015	.008								
	4	.970	.030	.015	6	.959	.041	.021								
	5	.948	.052	.026	7	.935	.065	.033								
	6	.918	.082	.041	8	.907	.093	.047								
	7	.874	.126	.063	9	.868	.132	.066								
7	2	.995	.005	.003	4	.992	.008	.004	5	.993	.007	.004				
	3	.990	.010	.005	5	.986	.014	.007	6	.989	.011	.006				
	6	.952	.048	.024	7	.965	.035	.018	9	.962	.038	.019				
	7	.927	.073	.037	8	.949	.051	.026	10	.947	.053	.027				
	8	.894	.106	.053	9	.927	.073	.037	12	.903	.097	.049				
					10	.899	.101	.051	13	.872	.128	.064				
8	3	.994	.006	.003	5	.992	.008	.004	7	.991	.009	.005	8	.993	.007	.004
	4	.989	.011	.005	6	.987	.013	.006	8	.986	.014	.007	9	.990	.010	.005
	7	.955	.045	.023	9	.957	.043	.021	11	.960	.040	.020	14	.950	.050	.025
	8	.935	.065	.033	10	.941	.059	.030	12	.946	.054	.027	15	.935	.065	.033
	9	.907	.093	.047	11	.919	.081	.041	14	.906	.094	.047	16	.917	.083	.042
	10	.873	.127	.064	12	.892	.108	.054	15	.879	.121	.060	17	.895	.105	.052
9	4	.993	.007	.004	6	.992	.008	.004	8	.992	.008	.004	10	.992	.008	.004
	5	.988	.012	.006	7	.988	.012	.006	9	.988	.012	.006	11	.989	.011	.006
	8	.958	.042	.021	11	.950	.050	.025	13	.958	.042	.021	16	.954	.046	.023
	9	.940	.060	.030	12	.934	.066	.033	14	.945	.055	.027	17	.941	.059	.030
	10	.917	.083	.042	13	.912	.088	.044	16	.909	.091	.045	19	.907	.093	.046
	11	.888	.112	.056	14	.887	.113	.057	17	.886	.114	.057	20	.886	.114	.057
10	5	.992	.008	.004	7	.993	.007	.004	10	.990	.010	.005	12	.991	.009	.004
	6	.987	.013	.006	8	.989	.011	.006	11	.986	.014	.007	13	.988	.012	.006
	9	.960	.040	.020	12	.958	.042	.021	15	.957	.043	.022	18	.957	.043	.022
	10	.945	.055	.028	13	.944	.056	.028	16	.945	.055	.028	19	.945	.055	.027
	12	.901	.099	.050	15	.907	.093	.047	18	.912	.088	.044	21	.917	.083	.042
	13	.871	.129	.065	16	.882	.118	.059	19	.891	.109	.054	22	.899	.101	.051
11	6	.991	.009	.004	8	.993	.007	.004	11	.992	.008	.004	14	.991	.009	.005
	7	.987	.013	.007	9	.990	.010	.005	12	.989	.011	.006	15	.988	.012	.006
	10	.962	.038	.019	14	.952	.048	.024	17	.956	.044	.022	20	.959	.041	.020
	11	.948	.052	.026	15	.938	.062	.031	18	.944	.056	.028	21	.949	.051	.025
	13	.910	.090	.045	17	.902	.098	.049	20	.915	.085	.043	24	.909	.091	.045
	14	.885	.115	.058	18	.878	.122	.061	21	.896	.104	.052	25	.891	.109	.054
12	7	.991	.009	.005	10	.990	.010	.005	13	.990	.010	.005	16	.990	.010	.005
	8	.986	.014	.007	11	.987	.013	.007	14	.987	.013	.007	17	.988	.012	.006
	12	.952	.048	.024	15	.959	.041	.021	19	.955	.045	.023	23	.953	.047	.024
	13	.936	.064	.032	16	.947	.053	.026	20	.944	.056	.028	24	.943	.057	.029
	14	.918	.082	.041	18	.917	.083	.042	22	.917	.083	.042	27	.902	.098	.049
	15	.896	.104	.052	19	.898	.102	.051	23	.900	.100	.050	28	.885	.115	.058

Table G cont'd.

	Size of Smaller Sample															
	9				10				11				12			
Size of Larger Sample	d	γ	α''	α'	d	γ	α''	α'	d	γ	α''	α'	d	γ	α''	α'
9	12	.992	.008	.004												
	13	.989	.011	.005												
	18	.960	.040	.020												
	19	.950	.050	.025												
	22	.906	.094	.047												
	23	.887	.113	.057												
10	14	.992	.008	.004	17	.991	.009	.005								
	15	.990	.010	.005	18	.989	.011	.006								
	21	.957	.043	.022	24	.957	.043	.022								
	22	.947	.053	.027	25	.948	.052	.026								
	25	.905	.095	.047	28	.911	.089	.045								
	26	.887	.113	.056	29	.895	.105	.053								
11	17	.990	.010	.005	19	.992	.008	.004	22	.992	.008	.004				
	18	.988	.012	.006	20	.990	.010	.005	23	.989	.011	.005				
	24	.954	.046	.023	27	.957	.043	.022	31	.953	.047	.024				
	25	.944	.056	.028	28	.949	.051	.026	32	.944	.056	.028				
	28	.905	.095	.048	32	.901	.099	.049	35	.912	.088	.044				
	29	.888	.112	.056	33	.886	.114	.057	36	.899	.101	.051				
12	19	.991	.009	.005	22	.991	.009	.005	25	.991	.009	.004	28	.992	.008	.004
	20	.988	.012	.006	23	.989	.011	.006	26	.989	.011	.005	29	.990	.010	.005
	27	.951	.049	.025	30	.957	.043	.021	34	.956	.044	.022	38	.955	.045	.023
	28	.942	.058	.029	31	.950	.050	.025	35	.949	.051	.026	39	.948	.052	.026
	31	.905	.095	.048	35	.907	.093	.047	39	.909	.091	.045	43	.911	.089	.044
	32	.889	.111	.056	36	.893	.107	.054	40	.896	.104	.052	44	.899	.101	.050

TABLE H t-Distribution

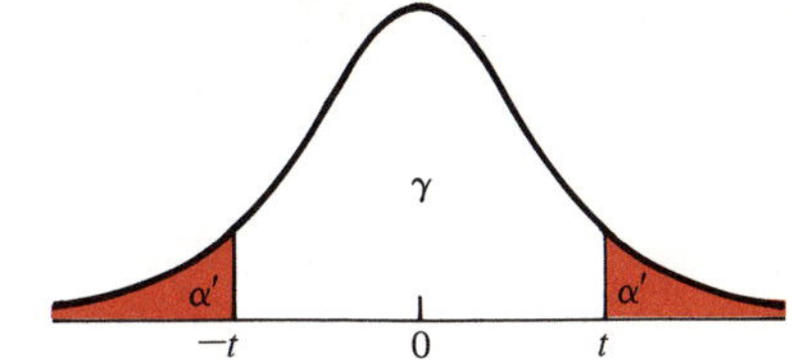

γ	.5	.8	.9	.95	.98	.99
α'	.25	.1	.05	.025	.01	.005
α''	.5	.2	.1	.05	.02	.01
df						
1	1.000	3.078	6.314	12.706	31.821	63.657
2	.816	1.886	2.920	4.303	6.965	9.925
3	.765	1.638	2.353	3.182	4.541	5.841
4	.741	1.533	2.132	2.776	3.747	4.604
5	.727	1.476	2.015	2.571	3.365	4.032
6	.718	1.440	1.943	2.447	3.143	3.707
7	.711	1.415	1.895	2.365	2.998	3.499
8	.706	1.397	1.860	2.306	2.896	3.355
9	.703	1.383	1.833	2.262	2.821	3.250
10	.700	1.372	1.812	2.228	2.764	3.169
11	.697	1.363	1.796	2.201	2.718	3.106
12	.695	1.356	1.782	2.179	2.681	3.055
13	.694	1.350	1.771	2.160	2.650	3.012
14	.692	1.345	1.761	2.145	2.624	2.977
15	.691	1.341	1.753	2.131	2.602	2.947
16	.690	1.337	1.746	2.120	2.583	2.921
17	.689	1.333	1.740	2.110	2.567	2.898
18	.688	1.330	1.734	2.101	2.552	2.878
19	.688	1.328	1.729	2.093	2.539	2.861
20	.687	1.325	1.725	2.086	2.528	2.845
21	.686	1.323	1.721	2.080	2.518	2.831
22	.686	1.321	1.717	2.074	2.508	2.819
23	.685	1.319	1.714	2.069	2.500	2.807
24	.685	1.318	1.711	2.064	2.492	2.797
25	.684	1.316	1.708	2.060	2.485	2.787
26	.684	1.315	1.706	2.056	2.479	2.779
27	.684	1.314	1.703	2.052	2.473	2.771
28	.683	1.313	1.701	2.048	2.467	2.763
29	.683	1.311	1.699	2.045	2.462	2.756
30	.683	1.310	1.697	2.042	2.457	2.750
40	.681	1.303	1.684	2.021	2.423	2.704
60	.679	1.296	1.671	2.000	2.390	2.660
120	.677	1.289	1.658	1.980	2.358	2.617
∞	.674	1.282	1.645	1.960	2.326	2.576

γ = area between $-t$ and t
= confidence coefficient

$\alpha' = \frac{1}{2}(1 - \gamma)$
= area above t
= area below $-t$
= significance level of one-sided test

$\alpha'' = 1 - \gamma = 2\alpha'$
= area beyond $-t$ and t
= significance level of two-sided test

TABLE I *F*-Distribution

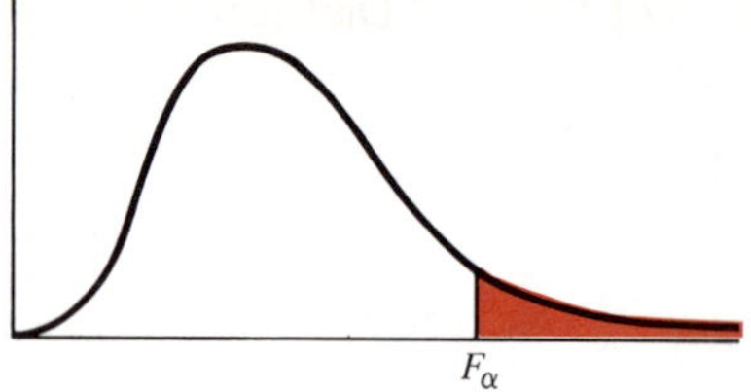

Significance Level $\alpha = .10$

Numerator Degrees of Freedom (columns); *Denominator Degrees of Freedom* (rows)

	1	2	3	4	5	6	7	8	9	10
1	39.86	49.50	53.59	55.83	57.24	58.20	58.91	59.44	59.86	60.19
2	8.53	9.00	9.16	9.24	9.29	9.33	9.35	9.37	9.38	9.39
3	5.54	5.46	5.39	5.34	5.31	5.28	5.27	5.25	5.24	5.23
4	4.54	4.32	4.19	4.11	4.05	4.01	3.98	3.95	3.94	3.92
5	4.06	3.78	3.62	3.52	3.45	3.40	3.37	3.34	3.32	3.30
6	3.78	3.46	3.29	3.18	3.11	3.05	3.01	2.98	2.96	2.94
7	3.59	3.26	3.07	2.96	2.88	2.83	2.78	2.75	2.72	2.70
8	3.46	3.11	2.92	2.81	2.73	2.67	2.62	2.59	2.56	2.54
9	3.36	3.01	2.81	2.69	2.61	2.55	2.51	2.47	2.44	2.42
10	3.29	2.92	2.73	2.61	2.52	2.46	2.41	2.38	2.35	2.32
11	3.23	2.86	2.66	2.54	2.45	2.39	2.34	2.30	2.27	2.25
12	3.18	2.81	2.61	2.48	2.39	2.33	2.28	2.24	2.21	2.19
13	3.14	2.76	2.56	2.43	2.35	2.28	2.23	2.20	2.16	2.14
14	3.10	2.73	2.52	2.39	2.31	2.24	2.19	2.15	2.12	2.10
15	3.07	2.70	2.49	2.36	2.27	2.21	2.16	2.12	2.09	2.06
16	3.05	2.67	2.46	2.33	2.24	2.18	2.13	2.09	2.06	2.03
17	3.03	2.64	2.44	2.31	2.22	2.15	2.10	2.06	2.03	2.00
18	3.01	2.62	2.42	2.29	2.20	2.13	2.08	2.04	2.00	1.98
19	2.99	2.61	2.40	2.27	2.18	2.11	2.06	2.02	1.98	1.96
20	2.97	2.59	2.38	2.25	2.16	2.09	2.04	2.00	1.96	1.94
21	2.96	2.57	2.36	2.23	2.14	2.08	2.02	1.98	1.95	1.92
22	2.95	2.56	2.35	2.22	2.13	2.06	2.01	1.97	1.93	1.90
23	2.94	2.55	2.34	2.21	2.11	2.05	1.99	1.95	1.92	1.89
24	2.93	2.54	2.33	2.19	2.10	2.04	1.98	1.94	1.91	1.88
25	2.92	2.53	2.32	2.18	2.09	2.02	1.97	1.93	1.89	1.87
26	2.91	2.52	2.31	2.17	2.08	2.01	1.96	1.92	1.88	1.86
27	2.90	2.51	2.30	2.17	2.07	2.00	1.95	1.91	1.87	1.85
28	2.89	2.50	2.29	2.16	2.06	2.00	1.94	1.90	1.87	1.84
29	2.89	2.50	2.28	2.15	2.06	1.99	1.93	1.89	1.86	1.83
30	2.88	2.49	2.28	2.14	2.05	1.98	1.93	1.88	1.85	1.82
40	2.84	2.44	2.23	2.09	2.00	1.93	1.87	1.83	1.79	1.76
60	2.79	2.39	2.18	2.04	1.95	1.87	1.82	1.77	1.74	1.71
120	2.75	2.35	2.13	1.99	1.90	1.82	1.77	1.72	1.68	1.65
∞	2.71	2.30	2.08	1.94	1.85	1.77	1.72	1.67	1.63	1.60

Significance level $\alpha = .05$

Numerator Degrees of Freedom

Denominator Degrees of Freedom	1	2	3	4	5	6	7	8	9	10
1	161.4	199.5	215.7	224.6	230.2	234.0	236.8	238.9	240.5	241.9
2	18.15	19.00	19.16	19.25	19.30	19.33	19.35	19.37	19.38	19.40
3	10.13	9.55	9.28	9.12	9.01	8.94	8.89	8.85	8.81	8.79
4	7.71	6.94	6.59	6.39	6.26	6.16	6.09	6.04	6.00	5.96
5	6.61	5.79	5.41	5.19	5.05	4.95	4.88	4.82	4.77	4.74
6	5.99	5.14	4.76	4.53	4.39	4.28	4.21	4.15	4.10	4.06
7	5.59	4.74	4.35	4.12	3.97	3.87	3.79	3.73	3.68	3.64
8	5.32	4.46	4.07	3.84	3.69	3.58	3.50	3.44	3.39	3.35
9	5.12	4.26	3.86	3.63	3.48	3.37	3.29	3.23	3.18	3.14
10	4.96	4.10	3.71	3.48	3.33	3.22	3.14	3.07	3.02	2.98
11	4.84	3.98	3.59	3.36	3.20	3.09	3.01	2.95	2.90	2.85
12	4.75	3.89	3.49	3.26	3.11	3.00	2.91	2.85	2.80	2.75
13	4.67	3.81	3.41	3.18	3.03	2.92	2.83	2.77	2.71	2.67
14	4.60	3.74	3.34	3.11	2.96	2.85	2.76	2.70	2.65	2.60
15	4.54	3.68	3.29	3.06	2.90	2.79	2.71	2.64	2.59	2.54
16	4.49	3.63	3.24	3.01	2.85	2.74	2.66	2.59	2.54	2.49
17	4.45	3.59	3.10	2.96	2.81	2.70	2.61	2.55	2.49	2.45
18	4.41	3.55	3.16	2.93	2.77	2.66	2.58	2.51	2.46	2.41
19	4.38	3.52	3.13	2.90	2.74	2.63	2.54	2.48	2.42	2.38
20	4.35	3.49	3.10	2.87	2.71	2.60	2.51	2.45	2.39	2.35
21	4.32	3.47	3.07	2.84	2.68	2.57	2.49	2.42	2.37	2.32
22	4.30	3.44	3.05	2.82	2.66	2.55	2.46	2.40	2.34	2.30
23	4.28	3.42	3.03	2.80	2.64	2.53	2.44	2.37	2.32	2.27
24	4.26	3.40	3.01	2.78	2.62	2.51	2.42	2.36	2.30	2.25
25	4.24	3.39	2.99	2.76	2.60	2.49	2.40	2.34	2.28	2.24
26	4.23	3.37	2.98	2.74	2.59	2.47	2.39	2.32	2.27	2.22
27	4.21	3.35	2.96	2.73	2.57	2.46	2.37	2.31	2.25	2.20
28	4.20	3.34	2.95	2.71	2.56	2.45	2.36	2.29	2.24	2.19
29	4.18	3.33	2.93	2.70	2.55	2.43	2.35	2.28	2.22	2.18
30	4.17	3.32	2.92	2.69	2.53	2.42	2.33	2.27	2.21	2.16
40	4.08	3.23	2.84	2.61	2.45	2.34	2.25	2.18	2.12	2.08
60	4.00	3.15	2.76	2.53	2.37	2.25	2.17	2.10	2.04	1.99
120	3.92	3.07	2.68	2.45	2.29	2.17	2.09	2.02	1.96	1.91
∞	3.84	3.00	2.60	2.37	2.21	2.10	2.01	1.94	1.88	1.83

Table I cont'd.

Significance level α = .01

Numerator Degrees of Freedom

Denominator Degrees of Freedom	1	2	3	4	5	6	7	8	9	10
1	4052	4999.5	5403	5625	5764	5859	5928	5982	6022	6056
2	98.50	99.00	99.17	99.25	99.30	99.33	99.36	99.37	99.39	99.4
3	34.12	30.82	29.46	28.71	28.24	27.91	27.67	27.49	27.35	27.2
4	21.20	18.00	16.69	15.98	15.52	15.21	14.98	14.80	14.66	14.5
5	16.26	13.27	12.06	11.39	10.97	10.67	10.46	10.29	10.16	10.0
6	13.75	10.92	9.78	9.15	8.75	8.47	8.26	8.10	7.98	7.8
7	12.25	9.55	8.45	7.85	7.46	7.19	6.99	6.84	6.72	6.6
8	11.26	8.65	7.59	7.01	6.63	6.37	6.18	6.03	5.91	5.8
9	10.56	8.02	6.99	6.42	6.06	5.80	5.61	5.47	5.35	5.2
10	10.04	7.56	6.55	5.99	5.64	5.39	5.20	5.06	4.94	4.8
11	9.65	7.21	6.22	5.67	5.32	5.07	4.89	4.74	4.63	4.5
12	9.33	6.93	5.95	5.41	5.06	4.82	4.64	4.50	4.39	4.3
13	9.07	6.70	5.74	5.21	4.86	4.62	4.44	4.30	4.19	4.1
14	8.86	6.51	5.56	5.04	4.69	4.46	4.28	4.14	4.03	3.9
15	8.68	6.36	5.42	4.89	4.56	4.32	4.14	4.00	3.89	3.8
16	8.53	6.23	5.29	4.77	4.44	4.20	4.03	3.89	3.78	3.6
17	8.40	6.11	5.18	4.67	4.34	4.10	3.93	3.79	3.68	3.5
18	8.29	6.01	5.09	4.58	4.25	4.01	3.84	3.71	3.60	3.5
19	8.18	5.93	5.01	4.50	4.17	3.94	3.77	3.63	3.52	3.4
20	8.10	5.85	4.94	4.43	4.10	3.87	3.70	3.56	3.46	3.3
21	8.02	5.78	4.87	4.37	4.04	3.81	3.64	3.51	3.40	3.3
22	7.95	5.72	4.82	4.31	3.99	3.76	3.59	3.45	3.35	3.2
23	7.88	5.66	4.76	4.26	3.94	3.71	3.54	3.41	3.30	3.2
24	7.82	5.61	4.72	4.22	3.90	3.67	3.50	3.36	3.26	3.1
25	7.77	5.57	4.68	4.18	3.85	3.63	3.46	3.32	3.22	3.1
26	7.72	5.53	4.64	4.14	3.82	3.59	3.42	3.29	3.18	3.0
27	7.68	5.49	4.60	4.11	3.78	3.56	3.39	3.26	3.15	3.0
28	7.64	5.45	4.57	4.07	3.75	3.53	3.36	3.23	3.12	3.0
29	7.60	5.42	4.54	4.04	3.73	3.50	3.33	3.20	3.09	3.0
30	7.56	5.39	4.51	4.02	3.70	3.47	3.30	3.17	3.07	2.9
40	7.31	5.18	4.31	3.83	3.51	3.29	3.12	2.99	2.89	2.8
60	7.08	4.98	4.13	3.65	3.34	3.12	2.95	2.82	2.72	2.6
120	6.85	4.79	3.95	3.48	3.17	2.96	2.79	2.66	2.56	2.4
∞	6.63	4.61	3.78	3.32	3.02	2.80	2.64	2.51	2.41	2.3

TABLE J **Random Digits**

47505	02008	20300	87188	42505	40294	04404	59286	95914	07191
13350	08414	64049	94377	91059	74531	56228	12307	87871	97064
33006	92690	69248	97443	38841	05051	33756	24736	43508	53566
55216	63886	06804	11861	30968	74515	40112	40432	18682	02845
21991	26228	14801	19192	45110	39937	81966	23258	99348	61219
71025	28212	10474	27522	16356	78456	46814	28975	01014	91458
65522	15242	84554	74560	26206	49520	65702	54193	25583	54745
27975	54923	90650	06170	99006	75651	77622	20491	53329	12452
07300	09704	36099	61577	34632	55176	87366	19968	33986	46445
54357	13689	19569	03814	47873	34086	28474	05131	46619	41499
00977	04481	42044	08649	83107	02423	46919	59586	58337	32280
13920	78761	12311	92808	71581	85251	11417	85252	61312	10266
08395	37043	37880	34172	80411	05181	58091	41269	22626	64799
46166	67206	01619	43769	91727	06149	17924	42628	57647	76936
87767	77607	03742	01613	83528	66251	75822	83058	97584	45401
29880	95288	21644	46587	11576	30568	56687	83239	76388	17857
36248	36666	14894	59273	04518	11307	67655	08566	51759	41795
12386	29656	30474	25964	10006	86382	46680	93060	52337	56034
52068	73801	52188	19491	76221	45685	95189	78577	36250	36082
41727	52171	56719	06054	34898	93990	89263	79180	39917	16122
49319	74580	57470	14600	22224	49028	93024	21414	90150	15686
88786	76963	12127	25014	91593	98208	27991	12539	14357	69512
84866	95202	43983	72655	89684	79005	85932	41627	87381	38832
11849	26482	20461	99450	21636	13337	55407	01897	75422	05205
54966	17594	57393	73267	87106	26849	68667	45791	87226	74412
10959	33349	80719	96751	25752	17133	32786	34368	77600	41809
22784	07783	35903	00091	73954	48706	83423	96286	90373	23372
86037	61791	33815	63968	70437	33124	50025	44367	98637	40870
80037	65089	85919	73491	36170	82988	52311	59180	37846	98028
72751	84359	15769	13615	70866	37007	74565	92781	37770	76451
18532	03874	66220	79050	66814	76341	42452	65365	07167	90134
22936	22058	49171	11027	07066	14606	11759	19942	21909	15031
66397	76510	81150	00704	94990	68204	07242	82922	65745	51503
89730	23272	65420	35091	16227	87024	56662	59110	11158	67508
81821	75323	96068	91724	94679	88062	13729	94152	59343	07352
94377	82554	53586	11432	08788	74053	98312	61732	91248	23673
68485	49991	53165	19865	30288	00467	98105	91483	89389	61991
07330	07184	86788	64577	47692	45031	36325	47029	27914	24905
10993	14930	35072	36429	26176	66205	07758	07982	33721	81319
20801	15178	64453	83357	21589	23153	60375	63305	37995	66275
79241	35347	66851	79247	57462	23893	16542	55775	06813	63512
43593	39555	97345	58494	52892	55080	19056	96192	61508	23165
29522	62713	33701	17186	15721	95018	76571	58615	35836	66260
88836	47290	67274	78362	84457	39181	17295	39626	82373	10883
65905	66253	91482	30689	81313	01343	37188	37756	04182	19376
44798	69371	07865	91756	42318	63601	53872	93610	44142	89830
35510	99139	32031	27925	03560	33806	85092	70436	94777	57963
50125	93223	64209	49714	73379	89975	38567	44316	60262	10777
25173	90038	63871	40418	23818	63250	05118	52700	92327	55449
68459	90094	44995	93718	83654	79311	18107	12557	09179	28408

TABLE J cont'd.

24810	61498	24935	86366	58262	44565	91426	86742	61747	79346
75555	42967	02810	16754	08813	40079	62385	21488	38665	94197
49718	05437	73341	17414	91868	24102	76123	67138	43728	43627
04194	83554	98004	14744	63132	75018	75167	24090	02458	78215
66155	24222	91229	63841	03271	56726	36817	51182	94336	20894
10801	76783	05312	30807	40006	04465	70163	17305	46414	76468
40422	75576	82884	27651	58495	87538	23570	66469	46900	95568
95572	12125	12054	17028	03599	73764	48694	85960	89763	58305
57276	49498	88937	08659	46840	83231	75611	94911	28467	67928
80566	94963	83787	87636	89511	64735	86699	66988	91224	72484
44517	55108	17435	33109	60343	46193	66019	43713	24097	52921
55424	87650	13896	90005	99458	20153	86688	13650	75201	79447
80506	78301	97762	16434	62430	28438	13602	63236	81431	75641
03646	54402	75413	39128	82975	73849	27269	73444	26120	06824
14537	53791	43951	51326	33274	54833	80802	66976	04878	35832
01644	33630	71247	59273	07811	33546	88628	06469	86257	39298
39387	94217	77995	53285	13354	84980	83590	63494	06036	18502
74962	49489	54662	93588	50466	55026	62458	06195	07995	71054
21165	45577	46383	38855	21561	89332	94248	09703	78397	38770
58519	95396	73607	72106	76597	85596	99075	39195	99605	66179
46982	79519	22294	15676	83484	98279	79200	02640	22501	43073
58463	67619	18006	05028	32441	83599	28915	05362	21612	64681
43055	00020	39254	68439	27399	24259	04641	50935	07112	55117
84073	38387	14337	90766	60436	65757	57590	17880	13776	35810
93542	37270	09361	62404	74056	52964	67372	81398	01482	97589
54467	20234	52813	85296	14542	73241	74848	39001	97598	76641
43608	42832	93917	67031	50220	94089	64858	27691	16719	99870
64808	01692	46424	64722	87162	06582	01452	14980	17397	07403
78703	93006	59651	48404	82284	66405	89818	00989	56112	78144
14886	70359	32158	30401	20829	22534	88848	07669	25100	48602
69280	61856	78974	91485	01583	11620	53740	32705	80391	56749
99680	99636	54107	79588	90845	21652	58875	13171	68531	18550
01662	21554	63836	41530	21864	81711	68921	61749	36051	78024
67852	69123	14280	17647	65125	82427	61594	32015	93473	05627
13911	67691	97854	89950	40963	06697	82660	69097	65284	49808
95822	09552	65950	34875	64250	41385	80133	70818	09286	30769
44068	24928	27345	34235	44124	06435	06281	43723	97380	76080
99222	66415	71069	62239	77467	35751	22548	23799	96272	58777
08442	61287	72421	35777	61079	42462	17761	94518	98114	74035
14967	60637	32097	28122	87708	19378	93372	23225	38453	80331
27864	15358	16499	91903	62987	98198	15036	23293	68241	44450
99678	16125	52978	79815	85990	18659	00113	93253	49186	25165
89143	79403	22324	54261	97830	42630	48494	09999	69961	39421
34135	31532	42025	83214	83730	28249	25629	11494	70726	45051
72117	97579	36071	29261	89937	78208	23747	56755	37453	51344
19725	76199	08620	22682	52907	25194	84597	93419	95762	14991
96997	66390	27609	41570	17749	23185	24475	56451	91471	33969
44158	67618	15572	95162	95842	08301	11906	68081	40436	58735
33839	40750	18898	61650	09970	47651	41205	65020	33537	01022
53070	61630	84434	05732	18094	71669	41033	82402	16415	83958

66558	78763	55932	15490	46790	47325	60903	15000	90970	06904
93810	69163	27172	10864	39108	79626	90431	44390	54290	70295
72045	47743	33163	88057	14136	55883	71449	68303	54093	95545
64251	86498	77947	21734	23571	86489	90017	24878	91985	03921
82220	31802	84619	51220	34654	60601	15088	26949	23013	72644
04991	91864	49269	66109	92609	37154	53225	73014	01890	04357
73895	65548	31996	73237	62411	22311	87875	79190	28237	73903
03515	01014	83955	11919	71533	71150	45699	95307	77713	66398
78808	89471	65152	62457	32410	14092	13813	08357	65485	83198
50648	45741	81584	54369	01575	92941	05484	41196	61946	89918
52074	39293	45087	07020	04753	66952	45199	83726	11602	57715
08209	01284	83775	89711	92322	31538	15808	94830	69581	94556
24292	33646	26925	04133	04895	07341	81441	53319	60118	98634
68189	39488	61468	23411	36471	65260	30134	55648	39176	61692
63096	33677	78900	30005	17324	83577	16699	62138	73469	89005
67106	05029	82711	17886	38351	42165	71101	37151	13547	38500
36272	89377	49623	71797	57532	90488	32967	60308	57256	66233
33143	08577	38507	85535	62784	29068	42392	41332	71636	49165
94138	78030	28934	91012	45780	66416	14003	05819	71031	00053
65199	99418	58039	96495	95954	48748	93022	46913	26250	35538
49897	14275	22123	84895	77567	17949	57969	31131	95882	08783
85679	52202	37950	09891	45369	48243	84985	08318	09853	86452
21441	75053	31373	89860	47671	33981	55424	33191	53223	54060
54269	06140	67385	01203	51078	48220	46715	09144	61587	63026
09323	22353	58095	97149	63325	18050	36840	26523	72376	89192
50214	26662	95722	78359	49612	75804	89378	96992	76621	91777
52022	22417	70460	91869	45732	54352	87239	41463	40310	45189
47176	39122	11478	75218	04888	49657	84540	49821	80806	83581
35402	23056	72903	95029	05373	35587	23297	11870	18495	79905
65519	29138	08384	41230	29209	87793	06285	90472	47054	57036
11259	05645	07492	16580	90016	22626	23187	25531	24281	05383
56007	97457	05913	74626	33923	25652	75099	73542	29669	82523
24558	05361	28136	58586	74390	95278	70229	15845	83717	91629
52889	89032	03429	81240	54824	16714	79590	91867	36732	16936
39534	82185	56489	40999	31361	98733	68769	77792	52694	14372
04948	01323	21617	23457	29217	65387	33130	30920	26298	84058
69496	38855	02249	50773	93315	41606	73918	32347	06673	95058
16068	36000	08084	66738	15982	82450	09060	49051	31759	54477
11907	18181	20687	05878	33617	16566	67893	50243	08352	64527
18382	61533	42865	90495	74809	70740	24939	43883	86674	40041
75960	66915	02595	66435	55610	42936	52336	15660	28110	10390
20156	11314	51105	46678	39660	54062	81972	55953	99513	41647
10528	02058	80359	99179	65642	93982	07133	39680	45791	79665
48514	83028	06720	33776	10023	14228	56367	04108	54855	77323
92644	48340	75864	50303	09037	02589	37463	81365	18567	65142
41477	25941	98283	49225	08721	84508	97549	46769	98389	19589
41630	05563	83127	71333	68606	49269	89244	53159	02762	18167
78489	08606	66190	02810	49460	63040	00221	27138	36645	02465
44971	55189	26570	98515	37222	54809	17570	64185	56333	72230
68023	51496	96693	76886	54420	59192	11645	54942	31693	28688

TABLE J cont'd.

79767	68246	17731	26380	53059	95517	78256	75888	07407	26606
51418	77326	50358	33736	42294	05839	41670	06190	79431	17649
99053	75365	68601	11974	06061	70547	34663	91460	16942	36190
76782	71815	25699	18820	46640	66131	68176	06721	96948	48831
00054	75974	39606	34585	11766	24425	59123	43770	01543	48199
76986	25299	56915	10445	17519	06383	69934	38270	50500	40036
09068	40312	32398	66316	86491	25159	68490	99215	13026	99583
14995	57799	04606	39602	96838	82575	60004	76945	87129	57982
52236	23601	02551	73803	43190	45217	25331	24319	47599	87713
04234	88134	76799	00690	00431	35795	68154	13726	92234	38523
09576	62666	62081	25900	85551	67305	74596	19856	95240	27096
08876	02110	88540	67911	22605	78936	53171	47839	31705	40128
90641	63802	62775	99910	54987	25337	87749	40698	13520	75360
81610	68893	76441	86754	68758	63132	70789	03593	60013	50974
24942	40614	82896	62376	22623	71501	62216	04926	92450	02354
69276	15411	39931	05682	84216	78727	45115	16172	48356	14454
26850	52243	72376	69365	46803	17763	81849	28183	15482	41846
58251	44226	99121	51709	55878	38127	62332	17779	60737	14139
80769	38995	78326	43032	42924	48921	92503	48883	39617	16926
77730	27199	41001	37730	17192	24001	68844	39054	03311	29616
14101	79409	55851	59065	09448	12362	37983	30132	44552	64721
78412	27673	22697	90409	80933	14397	65476	65039	54137	05837
80490	04572	83161	12758	77431	46984	36316	69058	80929	40091
48435	79357	07703	58519	22212	49369	39125	24281	10894	37515
55818	67243	32570	46300	30216	37922	59474	90998	32791	50913
38504	19642	20043	51907	12061	45989	34126	15466	60407	41060
62532	88768	10806	78056	56116	83894	04004	89425	98102	25472
41253	47292	21824	28796	66520	02299	35985	21745	67569	43758
21311	49354	26383	15911	58703	66264	52777	88468	78479	93144
79493	81187	22802	81892	23194	30874	29596	95527	05851	69557
86569	20549	41974	78598	04265	52585	06155	92795	47235	13421
71195	66679	83667	01622	90584	80413	10738	89237	40872	88433
34198	37972	11147	54830	11599	39138	63063	12883	50472	78611
49944	77066	51105	10926	02573	88622	25972	26116	20518	74519
20512	37842	93043	16725	38621	89067	28716	87520	86277	00223
95920	01676	50637	32471	85297	72345	36374	27257	66562	54449
69343	67580	74026	53503	36427	91961	40432	67921	77197	76739
63256	54571	48582	61218	68385	62550	00185	33937	14945	00532
07164	39622	15005	81428	35078	98992	58169	46310	02584	62127
32151	06034	35717	86386	09108	69625	80461	22900	75473	71302
36943	36628	02711	19174	95060	91057	91419	16271	92089	33761
42543	86517	62451	83617	33303	00232	91253	11758	42643	60941
82246	71725	68475	14115	88253	67433	08144	97231	14853	02076
13560	22591	42101	98738	59840	45639	67920	00794	88914	62635
13202	90236	87525	19154	01788	21129	30042	88748	86127	65421
06927	25724	01915	54331	39256	35530	93068	28035	38225	55304
06977	02761	45002	56046	62382	41952	37018	55770	84544	38292
71036	59813	93768	99757	94350	56118	52746	69784	08029	44572
52836	53177	39682	70280	84213	49817	79286	16600	47462	17033
27801	54044	16345	35740	82609	20754	88162	88875	02269	35975

TABLE K **Squares and Square Roots**

n	n^2	$\sqrt{n}$	$\sqrt{10n}$
1	1	1.000	3.162
2	4	1.414	4.472
3	9	1.732	5.477
4	16	2.000	6.325
5	25	2.236	7.071
6	36	2.449	7.746
7	49	2.646	8.367
8	64	2.828	8.944
9	81	3.000	9.487
10	100	3.162	10.000
11	121	3.317	10.488
12	144	3.464	10.954
13	169	3.606	11.402
14	196	3.742	11.832
15	225	3.873	12.247
16	256	4.000	12.649
17	289	4.123	13.038
18	324	4.243	13.416
19	361	4.359	13.784
20	400	4.472	14.142
21	441	4.583	14.491
22	484	4.690	14.832
23	529	4.796	15.166
24	576	4.899	15.492
25	625	5.000	15.811
26	676	5.099	16.125
27	729	5.196	16.432
28	784	5.292	16.733
29	841	5.385	17.029
30	900	5.477	17.321
31	961	5.568	17.607
32	1024	5.657	17.889
33	1089	5.745	18.166
34	1156	5.831	18.439
35	1225	5.916	18.708
36	1296	6.000	18.974
37	1369	6.083	19.235
38	1444	6.164	19.494
39	1521	6.245	19.748
40	1600	6.325	20.000
41	1681	6.403	20.248
42	1764	6.481	20.494
43	1849	6.557	20.736
44	1936	6.633	20.967
45	2025	6.708	21.213
46	2116	6.782	21.448
47	2209	6.856	21.679
48	2304	6.928	21.909
49	2401	7.000	22.136
50	2500	7.071	22.361
51	2601	7.141	22.583
52	2704	7.211	22.804
53	2809	7.280	23.022
54	2916	7.348	23.238
55	3025	7.416	23.452
56	3136	7.483	23.664
57	3249	7.550	23.875
58	3364	7.616	24.083
59	3481	7.681	24.290
60	3600	7.746	24.495
61	3721	7.810	24.698
62	3844	7.874	24.900
63	3969	7.937	25.100
64	4096	8.000	25.298
65	4225	8.062	25.495
66	4356	8.124	25.690
67	4489	8.185	25.884
68	4624	8.246	26.077
69	4761	8.307	26.268
70	4900	8.367	26.458
71	5041	8.426	26.646
72	5184	8.485	26.833
73	5329	8.544	27.019
74	5476	8.602	27.203
75	5625	8.660	27.386
76	5776	8.718	27.568
77	5929	8.775	27.749
78	6084	8.832	27.928
79	6241	8.888	28.107
80	6400	8.944	28.284
81	6561	9.000	28.460
82	6724	9.055	28.636
83	6889	9.110	28.810
84	7056	9.165	28.983
85	7225	9.220	29.155
86	7396	9.274	29.326
87	7569	9.327	29.496
88	7744	9.381	29.665
89	7921	9.434	29.833
90	8100	9.487	30.000
91	8281	9.539	30.166
92	8464	9.592	30.332
93	8649	9.644	30.496
94	8836	9.695	30.659
95	9025	9.747	30.822
96	9216	9.798	30.984
97	9409	9.849	31.145
98	9604	9.899	31.305
99	9801	9.950	31.464
100	10000	10.000	31.623

BIBLIOGRAPHY

General:

Bates, G. E. *Probability.* Addison-Wesley Publishing Company, Inc., Reading, Mass., 1965.

Campbell, S. K. *Flaws and Fallacies in Statistical Thinking.* Prentice-Hall, Inc., Englewood Cliffs, N.J., 1974.

Kruskal, W. K. "Statistics, Part 1." *International Encyclopedia of the Social Sciences.* (1968), 206–223. Macmillan & The Free Press, New York.

Moroney, M. J. *Facts from Figures.* Penguin Books, Baltimore, 1953.

Mosteller, F., editor. *Statistics by Example* (four volumes). Addison-Wesley Publishing Company, Inc., Reading, Mass., 1973.

Raiffa, H. *Decision Analysis.* Addison-Wesley Publishing Company, Inc., Reading, Mass., 1968.

Ruggles, R. and H. Brodie. "An Empirical Approach to Economic Intelligence in World War II." *Journal of the American Statistical Association.* 42 (1947), 72–91.

Tanur, J. M., editor. *Statistics, A Guide to the Unknown.* Holden-Day, Inc., San Francisco, 1972.

Wallis, W. A., and H. V. Roberts. *The Nature of Statistics.* The Free Press, New York, 1965.

Elementary Treatments of Probability and Statistics:

Blackwell, D. *Basic Statistics.* McGraw-Hill, Inc., New York, 1969.

Hodges, J. L., Jr., and E. L. Lehmann. *Basic Concepts of Probability and Statistics,* 2d ed. Holden-Day, San Francisco, 1970.

Mosteller, F., R. E. K. Rourke, and G. B. Thomas. *Probability and Statistics.* Addison-Wesley Publishing Company, Inc., Reading, Mass., 1961.

Wallis, W. A., and H. V. Roberts. *Statistics: A New Approach.* The Free Press, New York, 1956.

Statistical Methods:

Dixon, W. J., and F. J. Massey, Jr. *Introduction to Statistical Analysis.* 3d ed. McGraw-Hill, Inc., New York, 1969.

Natrella, M. G. *Experimental Statistics.* National Bureau of Standards Handbook 91, Superintendent of Documents, Washington, D.C., 1963.

Snedecor, G. W., and W. G. Cochran. *Statistical Methods,* 6th ed. Iowa State University Press, Ames, Iowa, 1967.

Statistical Tables:

Harvard University. *Tables of the Cumulative Binomial Probability Distribution,* 35. Annals of the Computation Laboratory, Cambridge, Mass.

Owen, D. B. *Handbook of Statistical Tables.* Addison-Wesley Publishing Company, Inc., Reading, Mass., 1962.

Pearson, E. S., and H. O. Hartley. *Biometrika Tables for Statisticians.* Vol. 1, 3d ed. Cambridge University Press, New York, 1967.

RAND Corporation, *A Million Random Digits with 100,000 Normal Deviates.* Free Press, Glencoe, Illinois, 1955.

The Dixon and Massey and the Natrella books contain many useful statistical tables.

ANSWERS TO SELECTED PROBLEMS

Answers are rounded to the number of decimals given.

Chapter 1

1a	14
1b	49.5
3a	81
3c	76
5	2
9	88, 69
12b	The extreme estimate is unchanged; the median estimate changes.
14a	155
14b	238
16	See Problem 18.
18a	See Problem 19.

Chapter 3

1a	$P(N) = .23$
1b	$P(Fr \text{ and } N) = .075$
1c	$P(L \text{ and } H) = .19$
1d	$P(N \text{ or } S) = .65$
1e	$P_{Se}(N) = .225$
3a	.125
5c	.366
8b	No
9	No
17a	$P_R(Q) = .080$, $P(Q) = .078$
20e	$.25 + .20 - .05 = .40$
22	.29

Chapter 4

1	$b(2) = .069$
3	$b(2) = 10p^2(1 - p)^3$
4	3
6c	.213
9a	.072
11b	.299
15	.198

Chapter 5

1a	.9394
2a	.1335
3a	.0753
4a	.3954
5a	.58
6a	.524
7a	1.645
8a	.3085
9a	.0003
11	17.2
12a	.59
15d	.954

Chapter 6

3	1024
4b	practically 1 for $\varepsilon = .10$
7a	(.5, .7)
9a	.28
15	No for $\gamma = .95$
16	Yes for $\gamma = .95$
18	9.6
20	14 percentage points
25	(.47, .60)

Chapter 7

2a	.09
2b	.10 for $p = .10$
4a	$p \neq .4$
4b	.06
6	.05
8a	210
9	.34
14b	.03
14e	.24
15	No; there are too many round thousands.
21	No

26	No
27	No; participants seem to be influenced by extraneous factors.

Chapter 8

1a	$p = \frac{1}{3}$
1b	$p > \frac{1}{3}$
2	$\delta' = .002$
5	$\delta'' = .04$
7	Neither is correct.

Chapter 9

2a	$\chi^2 = 3.5$ with 2 *df*
2b	$\chi^2 = 88.4$ with 8 *df*
4	Yes; almost too well.
9	Yes
11b	Probability computations assume independence.
12b	$\chi^2 = 2.6$

Chapter 10

1	$\chi^2 = 20.4$
2	$\chi^2 = 12.15$
4	$\chi^2 = 0.46$
9	No
14	$\chi^2 = 4.82$
17	Yes
18	News stories influence decisions of jurors.

Chapter 11

1	Person's age at death; age at which childhood disease occurs.
3b	$(a + b)/2$

Chapter 12

1a	S-interval: $(-1, 11)$ for $\gamma = .93$ W-interval: $(0.5, 10.5)$ for $\gamma = .945$
1c	$M = 6$; second estimate $= 5.5$
2	$S_- = 3$; $W_- = 7.5$
3a	Reject H if $S_+ < 12$.
8	Reject H: $\eta \le 75$ against A: $\eta > 75$ at level $\alpha = .047$ provided that $W_- < 22$. Test assumes symmetry of test scores.
9	No
10a	.081
10e	.03 using normal approximation
13a	488
13c	No
18b	$\delta'' = .75$ using the sign test
22	H: $\eta = 512$; A: $\eta > 512$; $S_- = 2$

Chapter 13

1	$W_+ = 1$; $\delta'' = .063$
5a	Accept H.
5b	W-interval: $(-1, 1.6)$ for $\gamma = .875$
5c	.15
9a	sign test statistic $= 18$ using $n = 44$ observations; $\delta'' > .174$
9b	sign test statistic $= 19$ using $n = 50$ observations; $\delta'' = .119$

Chapter 14

2b	3
2c	27
6	$(0.8, 2.5)$ for $\gamma = .959$
7	Yes; $U_x = 17.5$
11	Yes
14a	$U = 10$
14b	$(-2, 23)$ for $\gamma = .947$
16	No
18b	$\delta'' = .22$

Chapter 15

3	$H = 5.7$
4	$Q = 7.4$
5	test statistic $= 5.5$
6	test statistic $= 10.0$
7	test statistic $= 8.2$
9	test statistic $= 8.9$

10	test statistic = 6.0
21	Brand 1 is preferred to brand 3 (α = .10).
22	Method 3 produces higher scores than method 1 (α = .10).
23	Achievement level of mathematics majors is higher than that of sociology majors.

Chapter 16

2	$S = -18$; $\delta' = .013$
4	$S = 13$; $\delta' = .026$
10	$t = .80$
12	$t = .81$ in both cases
15a	Yes
20a	$r_S = .90$

Chapter 17

2a	49.5
2b	18.6
5c	$\bar{x} = 28712.5$; $M = 27800$
6c	3166
11	$t = 2.60$
12a	19.5
12b	(18.9, 20.1) for $\gamma = .95$
13	Accept H: $\mu \leq 30000$.
17a	Test statistic equals 24.1.

Chapter 18

4	(13.5, 24.5) for $\gamma = .95$
5b	1.58
5c	(0.87, 2.29) for $\gamma = .90$
7	test H: $\mu_y \geq \mu_x$ against A: $\mu_y < \mu_x$; $t = 1.53$.
8a	$t = -1.51$
8b	(−13.1, 3.1) for $\gamma = .98$
11	(0.93, 2.41) for $\gamma = .95$

Chapter 19

1	$MS_A = 237.53$; $MS_W = 62.98$; $F_a = 5.01$
4	Analysis of variance table:

Among	Within	Total
253.2	657.6	910.8
3	16	19
84.4	41.1	
	$F = 2.05$	

Chapter 20

1a	10; 20
1b	30; 65; 146
1c	5; 15; 46
1d	.978
3a	$m_{y.x} = 0.784x - 3.0$
3b	108.2
3d	.68
3f	$t = 3.49$
6a	$m_{x.y} = 38.8 + 0.52\,y$ (psychology: x, statistics: y)
6b	$m_{y.x} = 45.6 + 0.39x$
6d	$t = 1.44$
8a	.99
14	.735

Appendix

3	$E(R) = \frac{1}{6}(1 + 2 + \cdots + 6) = 7/2$
5	$V(R) = \frac{1}{6}(1^2 + 2^2 + \cdots + 6^2) - (7/2)^2$
7	See Example A.2.
11a	$E(I^2) = p \cdot 1^2 + (1 - p) \cdot 0^2 = p$

INDEX

DEFGHIJ–H–798